Systems Architecting

A *Business Perspective*

Systems Architecting

A *Business Perspective*

Gerrit Muller

CRC Press
Taylor & Francis Group
Boca Raton   London   New York

CRC Press is an imprint of the
Taylor & Francis Group, an **informa** business

CRC Press
Taylor & Francis Group
6000 Broken Sound Parkway NW, Suite 300
Boca Raton, FL 33487-2742

© 2012 by Taylor & Francis Group, LLC
CRC Press is an imprint of Taylor & Francis Group, an Informa business

No claim to original U.S. Government works

International Standard Book Number: 978-1-4398-4762-6 (Hardback)

Visit the Taylor & Francis Web site at
http://www.taylorandfrancis.com

and the CRC Press Web site at
http://www.crcpress.com

Systems Architecting: A Business Perspective

Gerrit Muller

Systems Architecting: A Business Perspective

LOC PAGE

Dedication

This book is dedicated to my wife Lia Charité
who left me free to travel, to teach, and to write,
who kept watch over my well-being,
and who helped me to reflect on the contents of the book.

Preface

USE IN EDUCATION

This book is written as a textbook so that it can be used for graduate- and postgraduate-level education. It can be used for courses ranging from introductory, for example, one credit, to full postgraduate courses of four credits.[1]

The provided exercises can be applied to a fictional company or on a real company. We recommend using a real company whenever the students have working experience at that company. Otherwise, the teacher identifies a product that will engage the students, and they then imagine that they are part of the management of the fictional company developing and delivering the product.

A teacher's guide is available to facilitate the selection of suitable subjects, explaining the background of exercises and the assessment of answers. All diagrams are available as slides on the supporting websites:

- http://www.crcpress.com/product/isbn/9781439847626
- http://www.gaudisite.nl/SABP.html

Slides and diagrams can only be used if a reference to the source is presented.

INTERMEZZOS

Systems architecting methods and techniques always have to be adapted to the business, the market, the organization, and the technology. The same holds true for the education of systems architecting. The core of the book can be shared for most curriculums. In the *intermezzos* we discuss additional subjects that can be used by the teacher when appropriate.

Most intermezzos describe subjects that are hot discussion topics in practice. For example, 2.2 discusses the different titles and terms used in practice for the same or similar roles. Knowledge of all different titles and terms is not mandatory to become a systems architect. However, clarifying the titles and roles may improve communication between people with different backgrounds.

THE EXPLICIT STRUCTURE OF THIS BOOK

We address in this book the contribution of systems architecting to the business. The approach we take is similar to the approach taken by architects in their work: the so-called "viewpoint hopping." Figure 1 shows how we will move from viewpoint to viewpoint.

---

[1] we use US credits as unit. Expressed in European credits (ECTS), the study load can be varied from 2 to 10 ECTS

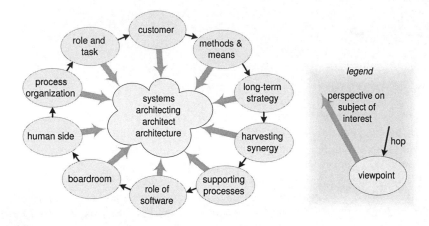

Figure 1 Viewpoint hopping applied to systems architecting.

We look at systems architecting from different viewpoints to create a better understanding of why, what, how, and when systems architecting contributes to the business.

Session 1 Process and Organization
Session 2 Role and Task of the Systems Architect
Session 3 From Customer Understanding to Requirements
Session 4 Systems Architect Methods and Means
Session 5 Long-Term Strategy
Session 6 Harvesting Synergy, Product Families
Session 7 Supporting Processes
Session 8 The Role of Software in Complex Systems
Session 9 Boardroom presentation
Session 10 Human side
Session 11 Wrap-up, Expectations, How to continue, Evaluation

Figure 2 The full course program covered by this book.

Figure 2 shows the same viewpoints that a full course based on this book would address, mapped on the sessions of the course. There is a chapter dedicated to each session.

Architects in their daily work apply viewpoint hopping much faster than in the course. Architects will also iterate many times over all viewpoints. Improved understanding of one viewpoint impacts the understanding of other viewpoints. We recommend that students iterate over the material several times. Exercises and project work are designed to facilitate such iteration.

THE VALUE AND LIMITS OF SYSTEMS ARCHITECTING

We will see that systems architecting is an integrating discipline that relates to all technical and business disciplines. The core value of systems architecting is that it facilitates the creation of an appropriate solution to the right problem.

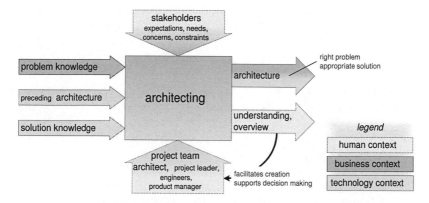

Figure 3   Architecting helps to find an appropriate solution for the right problem.

Insufficient understanding of the given problem often leads to project teams creating a solution to the wrong problem. Insufficient understanding of the solutions causes the selection of less optimal realizations. The core of systems architecting is to understand the problem and solution to such an extent that the program and project teams have sufficient overview and insight to make appropriate decisions, see Figure 3.

Project management, marketing management, and engineering are complementary core disciplines in product creation. **Systems architecting is not a silver bullet** that will resolve all problems in product creation. Without the other core disciplines, it is unable to create any system at all. However, good systems architecting improves the result and decreases problems, such as project delays, cost overruns, and poor performance of the system.

RED THREADS

Recurring principles or fundamental concepts to systems architecting are woven through the chapters in the book as *red threads*:

Communication between many stakeholders with different backgrounds. Systems architecting involves lots of communication. Facilitation of communication is one of the core values of systems architecting in the organization.

Understanding of problems and solutions to make appropriate choices. Systems architects have the urge to understand what is needed or how a solution really works.

Providing overview and insight to all stakeholders. Most stakeholders see only a small part of problem and solution. Systems architects empower other stakeholders by providing insight in the context.

Awareness of unknowns and uncertainties in all aspects of systems architecture. Systems architects operate in an environment full of unknowns and uncertainties. They need to provide direction despite the lack of certainty.

Goals and means articulation. Quite often, people confuse goals and means, causing the means to be chased rather than the goals. Architects need to disentangle goals and means, where often the means at one level can be the goal at the next level.

Customer and life-cycle context . A major step in the development from designer to architect is the inclusion of customer and life-cycle context in the daily work. The technical education of most designers does not typically include or promote this contextual knowledge. In practice, the extension of the scope is stretching the course participants significantly.

Being sharp and factual rather than acting on beliefs. Architects need to dive in to get facts and figures. Modeling and analysis turns these facts and figures into meaningful information that forms the foundation for decisions.

Feedback and iteration as leading principles in doing architecture work. Feedback is a mechanism to ensure that the work is heading toward the intended goal. Iteration is a way to frequently include the many different viewpoints that determine the architecture.

The above description is terse. We recommend that students regularly reflect on the learning material using this list.

THE Gaudí PROJECT

• Consolidate existing Systems Architecting Methods

evaluate, reflect, generalize

• Make the Systems Architecting art more accessible

case descriptions

• Enable the education of (future) System Architects

curriculum, course material

• Research new or improved Systems Architecting Methods

industry as laboratory

Figure 4   Goals of the Gaudí project.

This book is based on material from the Gaudí project. It started in 1999 with the goals shown in Figure 4. The project is named after the famous Spanish architect Antoni Gaudí (1852-1926).

The philosophy of the project is that slides, papers, books, and courses improve in quality by obtaining feedback. All material created as part of the Gaudí project is made public very early in its life cycle; see www.gaudisite.nl. Early exposure facilitates early feedback. The material is also copyrighted such that it can be reused freely; the only prerequisite for reuse is that the source of the material be mentioned.

Transforming some of the material and courses into a book is the last step in maturing the material. The book and Gaudisite.nl are complementary. Teachers and readers are invited to explore the website, where an order of magnitude more architecting-related material can be found.

FEEDBACK

Feedback on the book, the associated teacher's manual, and the supporting material is welcome. Feedback can be sent by email to gerrit.muller@gmail.com.

ACKNOWLEDGMENTS

Many people have contributed to this book directly or indirectly during the past three decades, among them colleagues, employers, and friends.

The Gaudí project started at Philips Research. The management supported this alternate way of working, and Jaap van der Heijden paved the way. Henk Obbink provided insights and challenges. Many colleagues at Philips Research either contributed or provided feedback.

The primary material of the Gaudí project is based on architecting work at previous employers, especially Philips Health Care and ASML. Managers such as Hans Brouwhuis enabled the project by permitting publication of cases. Colleagues from that period, such as Ben Spierenburg, Jan Statius Muller, and Maarten Bonnema, provided inputs and feedback.

The Philips Training department (CTT) facilitated the SARCH (Systems Architecting) training for many years. Vincent Ronteltap explained the didactic aspects of such training. Paul de Witte supported the development of this course over the years. Other teachers, such as Pierre America and Ger Schoeber, provided feedback on the material and the book.

The Gaudí project moved from Philips Research to the Embedded Systems Institute (ESI) in Eindhoven, the Netherlands. Martin Rem, scientific director of ESI, was inspirational and helped me to improve writing in English. ESI management supported the hosting of the website and the further development of the content. Many colleagues contributed or provided feedback. Especially, David Watts and Roland Mathijssen were instrumental in creating the book.

Since 2008, Buskerud University College has been supporting the Gaudí project. Gunnar Berge has been a great supporter of this book. Merete Faanes introduced Reflection as part of Systems education.

Many people worldwide have sent reactions on the Gaudí material to me. Thomas McKendree asked his class of Systems Architecting students to review papers on the website and to send feedback to me. Jonathan Losk stimulated writing the book and reviewed parts of it.

The Systems Engineering community, embodied in INCOSE, was inspirational. Discussions with highly experienced people like Tom Gilb and Bud Lawson helped to get perspective. The dean of Systems Engineering at Stevens Institute, Dinesh Verma, has been supportive.

Global colleagues, especially Ruth Malan and Dana Bredemeyer, stimulated the book effort.

The editor at CRC Press, Leong Li-Ming, has been inspiring and helpful.

Finally, I would like to thank all the numerous readers, course participants, and students who asked their questions and described their cases. They formed a tremendous source of knowledge, insights, and inspiration.

Contents

1 Process and Organization

1.1 PROCESS DECOMPOSITION OF A BUSINESS

1.1.1 INTRODUCTION

This chapter positions the system architecting process in a wider business scope. The objective of this chapter is to provide systems architects insight into the business processes, and especially into the processes to which systems architects actively contribute.

The focus is on companies that create physical products. Other types of businesses, such as solution providers and providers of services and courseware, also need systems architecting. The process structure will deviate somewhat from the structure presented here. See Section 1.6 Intermezzo: "Products, Projects, and Services," for a discussion on the processes in these other businesses.

1.1.2 PROCESS DECOMPOSITION

The business process can be decomposed into four main processes as shown in Figure 1.1. We have intentionally ignored the supporting and connecting processes. This simplification will allow us to get a number of more fundamental insights into the main processes.

The four main processes are described below:

Customer-Oriented Process performs in repetitive mode all direct interaction with the customer. This process is the cash-flow-generating part of the enterprise. All other processes only spend money.

Product Creation Process feeds the Customer-Oriented Process with new products. This process ensures the continuity of the enterprise by creating products that keep the company competitive. In this way, the Product Creation Process enables the Customer Oriented Process to generate cash flow in the near term as well.

People, Process, and Technology Management Process manages the competencies of the employees and the company as a whole. The competencies of the employees and the company are the main assets of a company.

Policy and Planning Process is the management process. The Policy and Planning Process defines the strategy; the long-term direction of the company, and it balances the shorter-term tensions between the three other main processes. The Policy and Planning Process uses roadmaps and budgets to define the direction for the other processes. Roadmaps give direction to the Product Creation Process and to the People, Process, and Technology Management Process. These roadmaps are transformed into budgets and plans for the medium term, which are mandatory for all stakeholders.

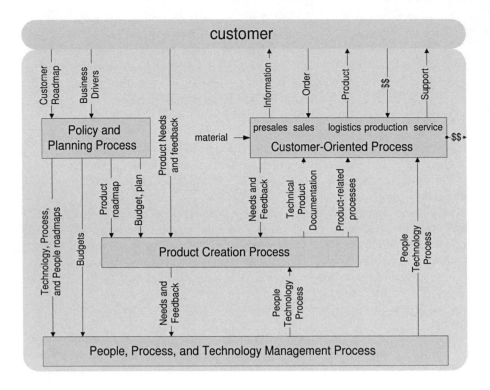

Figure 1.1 Simplified decomposition of the business in four main processes.

Figure 1.2 shows a financial perspective on these four processes. The financial terms illustrate the different time scales of these processes. The Customer-Oriented Process generates cashflow; that makes it an urgent process. The Product Creation Process ensures cash flow for the near future; this process is important for the mid-term. The People, Process, and Technology Management Process manages the assets of the company; a long-term consideration.

The four processes as described here are different in nature. The Customer-Oriented Process executes over and over a well-defined set of activities. The systems architect does not participate in an active role in this process. However, since the Customer-Oriented Process is the main customer of the Product Creation Process, it is crucial that the systems architect understands, or even better, has experienced, the Customer-Oriented Process.

The systems architect is in continuous interaction with many stakeholders, mostly about technical aspects. From this perspective, the architect will generate inputs for the People, Process, and Technology Management Process. This might result in participation in this process, for instance, by coaching, participation in the appraisal process, or participation in technology studies.

The number of instances of each process is related to different entities:

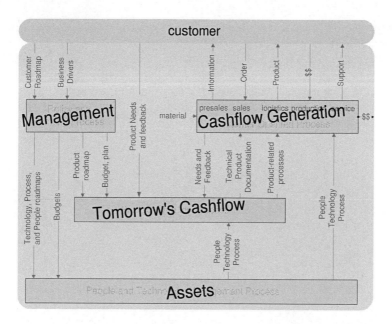

Figure 1.2 Decomposition of the business into four main processes, characterized by their financial meaning.

Customer-Oriented Process: Depends on geography, customer base, and supply chain.

Product Creation Process: One per entity to be developed, where such an entity can be a product family, a product, or a subsystem.

People, Process, and Technology Management Process: One per "competence," where a competence is a cohesive set of technologies and methods.

Policy and Planning Process: One per business. This is the proactive integrating process.

The evolutionary developments of product variants and new releases are seen as individual instances of the Product Creation Process. For example, the development of a single new feature for an existing product is performed by following the entire Product Creation Process. Some steps in the process might be (nearly) empty, which does not cause any harm.

1.1.3 PROCESS VERSUS ORGANIZATION

This process decomposition is not an organization; see Section 1.2, Intermezzo: "What is a Process?". A single person can (and often will) fulfill several roles in different processes.

Systems architects specifically spend most of their time in the *Product Creation Process* (about 75%), a considerable amount of time in the *Policy and Planning*

Process (about 20%), and a small fraction of their time in the *People, Process, and Technology Management Process*.

Most engineers will spend a small amount of time in the *People, Process, and Technology Management Process*, working on technologies and capabilities, while the majority of their time is spend in the *Product Creation Process*.

1.1.4   VALUE CHAIN AND FEEDBACK

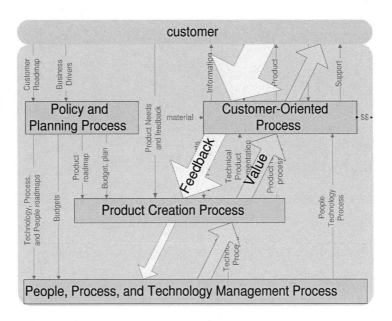

Figure 1.3   The value chain and the feedback flow in opposite direction.

The value chain in these processes starts at the assets in the People, Process, and Technology Management Process. The assets are transformed into potential money by the Product Creation Process. The Customer-Oriented Process finally turns it into real money. Figure 1.3 shows the value chain.

The feedback flows in the opposite direction, from the customer via the *Customer-Oriented Process* and the *Product Creation Process* to the *People, Process, and Technology Management Process*. Customers will communicate mostly with sales and service people. Needs and complaints are filtered by the reporting system before the information reaches the Product Creation Teams. Only a small part of the customer feedback reaches the People, Process, and Technology Management Process.

This simple model explains why the knowledge about the customer gets less when we get deep into the organization. The consequence is that internal technology and process shows too little concern for urgent customer or business challenges; the sense of urgency seems to be lacking. We can take preventive measures, such as sending process and technology managers to customer sites once we are aware of the gap caused by this natural information flow.

1.1.5 DECOMPOSITION OF THE CUSTOMER-ORIENTED PROCESS

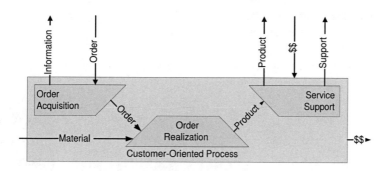

Figure 1.4 Decomposition of the Customer-Oriented Process.

The Customer-Oriented Process is often the largest process in terms of money. From the business point of view, it is an oversimplification to model this as one monolithic process. Figure 1.4 shows a further decomposition of this process.

The *Order Acquisition Process* and the *Service Support Process* are operating quite close to the customer. The *Order Realization Process* is already somewhat distant from the customer.

The owners of all these three processes are stakeholders of the *Product Creation Process*. Note that these owners have different interests and different characteristics.

1.2 INTERMEZZO: WHAT IS A PROCESS?

1.2.1 INTRODUCTION

We rely in this chapter heavily on the notion of a process. This intermezzo is defining "process" for the context of this book. We define "process" since this word is heavily overloaded in our daily world. We also discuss the relationship of processes with organizations and the drive for process improvement.

1.2.2 WHAT IS A PROCESS?

We use process as an abstracted way of working. A process can be characterized by the following attributes:

Purpose *What is to be achieved and why?*
Structure *How will the goal be achieved?*
Rationale *What is the reasoning behind this process?*
Roles *What roles are present, what responsibilities are associated, what incentives are present, what are the criteria for these roles?*

Ordering *What phasing or sequence is applied?*

In [11] the following definition is given:

A process is an activity that takes place over time and has a precise aim regarding the result to be achieved. The concept of a process is hierarchical, which means that a process may consist of a partially ordered set of subprocesses.

This definition parallels the characterization above. It adds explicitly the potential hierarchical decomposition of the process itself.

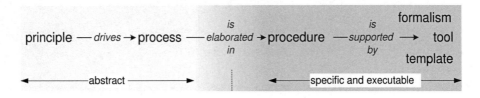

Figure 1.5   A process within an abstraction hierarchy.

The notion of a process can be seen as one step in an abstraction hierarchy, as shown in Figure 1.5. The most abstract notion in this hierarchy is the "principle." A principle is a generic insight that can be used for many different purposes. An example of a principle is decomposition: *Whenever we have something big, for example, a problem or project, then we can decompose it into smaller pieces. These smaller pieces are easier to solve or create than the original big one.*

A process is rather abstract. It describes the essentials of the purpose, structure, rationale, roles, and timing, leaving plenty of implementation freedom. The power of a process is its abstraction, which enables its application in a wide range of applications, by tailoring its implementation to the specific application.

A process can be tailored and elaborated in one or more procedures that describe, like a cookbook, what needs to be done when and by whom. The "why" in a procedure has often disappeared, to be replaced by practical information for the execution.

The implementation of a procedure can be supported by tools, formalisms or notations, templates, and other means.

In practice, managers and employees ask for tools (means) and procedures (what and how). However, without an understanding of the thinking behind the procedure (why) as given in the process, these tools and procedures can be meaningless. The process captures the rationale behind procedures, tools, notations, templates, and other means.

1.2.3   THE RELATION BETWEEN PROCESSES AND ORGANIZATIONS

Traditional management is focused on "organizations," where organizations are characterized by the attributes shown in Figure 1.6.

What **functions** are needed?

Who is **responsible** for this function?

What is the **hierarchical relation** between the functions?

What **meeting structure** is required?

Figure 1.6　Organizational attributes.

This management view is insufficient in today's fast-moving complex world. The weak spots of the organizational view are shown in Figure 1.7.

Many activities cut arbitrarily through the 1-dimensional hierarchy, causing

　　lack of ownership, unclear responsibilities

　　high impedance transitions at organizational boundaries

Functions are a combination of tasks, where, in most cases, no human exists with the required skills

Meeting structures are insufficient and inefficient to get things done

Figure 1.7　Weaknesses of the organizational view.

Processes are more modern instruments for management. Many processes are required to ensure the effective functioning of an organization. These processes are interrelated and overlapping. Processes are nonorthogonal and do not fit in a strict hierarchical structure.

Most complex product developments do not fit in the classical hierarchical organization model but require a much more dynamic organization model, such as the currently popular more chaotic network organization. Processes are the means that help to ensure the output of dynamic organization models such as a network organization.

Processes can be seen as the blueprint for the behavior of the people within the organization. People will fulfill multiple roles in multiple processes. The process description is intended to give them a hold on what is expected from them.

All important activities will be covered by a process, requiring the definition of ownership, relation with other processes, etc. The allocation of roles to people is much more dynamic than in conventional hierarchies. More dynamic allocation enables a better match between personal capabilities and required skills. In practice, dynamic allocation leads to more distribution of responsibilities, making it more feasible to match capabilities and skills.

The *80/20 rule is also valid for processes: 80% of the behavior is covered by the processes, while 20% requires independent creative behavior. An organization without processes drowns in chaos, while an organization that blindly implements them will be killed by its own inertia, its inability to adapt to the fast-changing world.*

For reasons of continuity and stability, a hierarchical organization is required. The slowest evolving dimension, such as the competence, technology, or application domain, is mostly used as a basis for this hierarchy. This hierarchy functions as an anchor point for people in the continuously changing process world, but should play only a minor role in the entire operation.

The **Centurion** *turnaround operation within Philips, orchestrated by CEO Jan Timmer in the early nineties, urged the Philips managers and employees to change from an introverted organization point of view to an external result-oriented process point of view.*

1.2.4 PROCESS IMPROVEMENT

Organizations look for ways to improve their efficiency, urged by competitive pressure . Many opportunities for improvement have a strong process component.

The 7S model by McKinsey gives a practical way to improve an organization in a balanced way. The message behind this model is that at least 7 views must be balanced when changing an organization. These 7 views are Shared value, Strategy, Systems, Style, Staff, Skills, and Structure.

The most common pitfall in improvement programs is the overemphasis on the process component, or worse, the isolation of the process improvement. Organizations assessing their maturity level, for instance by Maturity Models [20], quite often get too much process focus. The Process Improvement Officer[1] is focused on process issues only. Hence, where the process view is introduced as an extroverted result-oriented approach, it suddenly turns into an introverted improvement program, where business goals and drivers are unknown.

This is quite a sad situation: The opportunities for improvement are ample with a strong process component; however, due to the wrong focus, a negative effect is obtained (such as rigid procedures).

Recommendation: *Process improvements should originate from the directly involved people, for instance, project leaders, engineers, architects, etc. Invite participation by this group in such a way that its members that they own the process.*

1.3 PRODUCT CREATION PROCESS

1.3.1 INTRODUCTION

The *Product Creation Process* describes how an organization gets from a product idea to a tested system and all product documentation that is required for the

---

[1]The existence of this function in itself is a risk; it invites the unbalanced isolated "improvement" behavior

Customer-Oriented Process. Systems architects spend most of their time in the *Product Creation Process*. We will discuss the *Product Creation Process*, including organizational aspects and the roles of people within the process.

1.3.2 THE CONTEXT OF THE PRODUCT CREATION PROCESS

Figure 1.1 shows the context of the *Product Creation Process* in the decomposition of the business into four main processes. From the *Product Creation Process* point of view, the *Policy and Planning Process* determines the charter for the *Product Creation Process*. The *Technology, People, and Process Management Process* supplies people, process, and technology, enabling product creation. The *Customer-Oriented Process* is the customer: it receives and uses the results of product creation.

The *Product Creation Process* has a much wider context than the conventional "Research and Development" or "Development and Engineering" departments. The *Product Creation Process* includes everything that is needed to create a new product; for instance it includes

- Development of the production process
- Design of the logistics flow and structure
- Development of required services
- Market announcement
- Market introduction

In other words, the Product Creation Process is a synchronized effort of nearly all business disciplines within a company.

The term Product Creation is not only used for the development of entirely new products but also applies to the development of variations of existing products or the development of upgrades or add-on products. The implementation of the *Product Creation Process* can vary, depending on the product being developed; a small add-on product will use a different organization than the development of a large new complex product.

1.3.3 PHASES OF THE PRODUCT CREATION PROCESS

The Product Creation Process can be structured by using a phased approach. Figure 1.8 shows the phases as used in this book. The figure shows the participation of all business disciplines during this process.

These phases are used across all business functions that have to participate in the Product Creation Process. It is a means to manage the relations between these functions and to synchronize them. Note that sales, production, logistics, and service people are involved in the Product Creation Process. Their participation is required to understand the needs of the Customer-Oriented Process. A good understanding of these needs is required to develop the new procedures and processes for the Customer-Oriented Process, such as ordering, manufacturing, and installation.

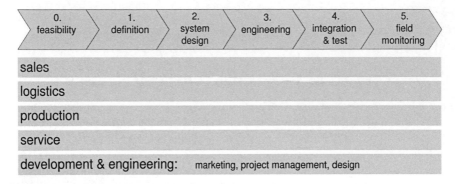

Figure 1.8 A phased approach to the Product Creation Process, showing the participation of all disciplines during the entire process.

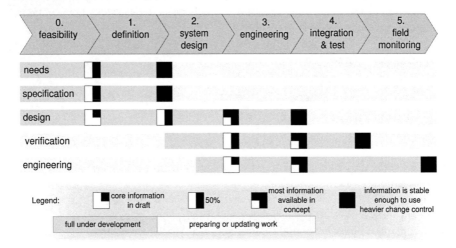

Figure 1.9 A phased approach to the Product Creation Process, showing the progress of the different design deliverables.

Figure 1.9 zooms in on the expected progress of the design deliverables. We use the term "workflows" for the horizontal classes of activities: *needs analysis*, *product specification*, *design*, *verification and validation*, and *engineering*. Note that *needs analysis*, *product specification*, and *design* progress concurrently. Also note that the first review typically takes place long before any of the workflows is complete. The main question for the first review is: does it make sense to invest in the later phases?

The advantages of a phased approach are shown in Figure 1.10. The project members get guidelines from the phase model: *who* does *what* and *when*. At the same time, the checklists per phase provide a means to check the progress for the management team. The main risk is the loss of common sense, where project members or management team apply the phase model too rigidly.

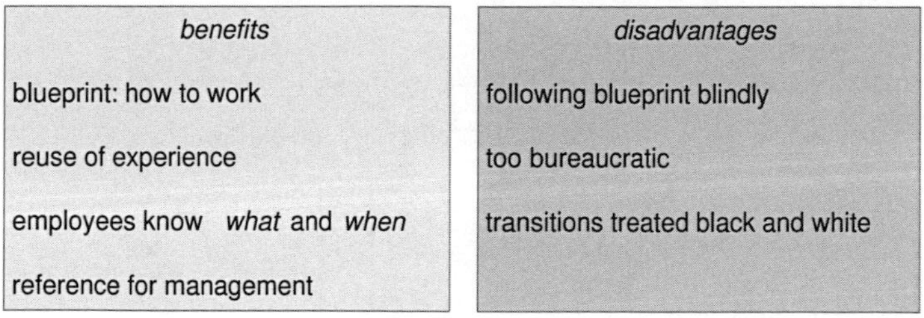

benefits	disadvantages
blueprint: how to work	following blueprint blindly
reuse of experience	too bureaucratic
employees know *what* and *when*	transitions treated black and white
reference for management	

Figure 1.10 Advantages and disadvantages of a phased approach.

Customization of the phase model to the specific circumstances is always needed. Keep in mind that a phased process is only a means.

The phase process is used as a means for the management team to judge the progress of the Product Creation Process. That can be done by comparing the actual progress with the checklists of the phase model at the moment of a phase transition. The actual progress is measured at the moment of transition. Normally, the development will continue after the phase review even if some deliverables are behind schedule. In that case, the problem is identified, enabling the project team to take corrective action. Some management teams misinterpret the phase transition as a milestone with mandatory deliverables. Based on this misinterpretation, the management team might demand full compliance with the checklist, disrupting the project. This kind of interference can be very counterproductive. See section 1.3.5 for a better management method with respect to milestones.

The important characteristics of a phase model are shown in Figure 1.11:

Concurrency of need analysis, specification, design, and engineering, and concurrency between activities within each of these workflows.
Checkpoints at phase transition. Often more checkpoints are defined, for instance, half way through a phase.
Iteration over the workflows and over activities within the workflows.
Large-impact decisions that have to be taken, often long before the full consequence of the decisions can be foreseen.

1.3.4 EVOLUTIONARY MODELS FOR PRODUCT CREATION

The phase model stresses and supports concurrent activities; see also [8]. A common pitfall is a waterfall interpretation of a phased approach. Following a strict top-down approach can be a very costly mistake because feedback from implementation and customers is, in that case, too late in the process. Early and continuous feedback, both from the implementation and customer point of view, is essential; see Section 1.4, Intermezzo: "The Importance of Feedback".

High market dynamics expose one weakness of the phased approach: market and user feedback becomes available at the end of the creation process. This is a sig-

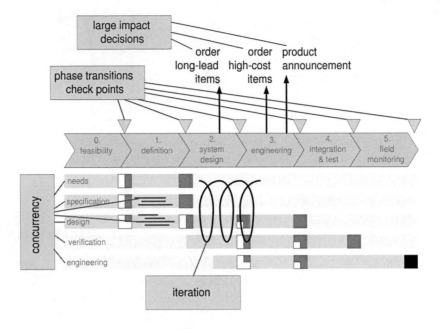

Figure 1.11   Characteristics of a phase model

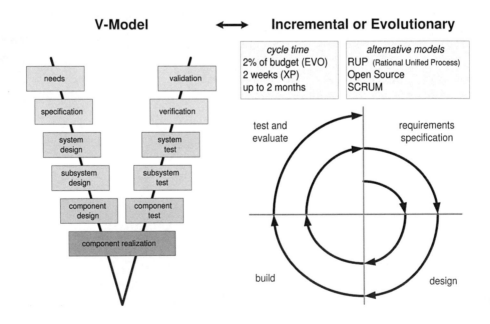

Figure 1.12   V-model versus incremental or evolutionary development models.

nificant problem because most product creations suffer from large uncertainties in the specifications. Discovering at the end that the specifications are based on wrong assumptions is very costly.

Figure 1.12 shows the V-model and evolutionary model side by side. Evolutionary methods focus on early feedback creation. EVO [7] by Gilb recommends using evolutionary development steps of 2% of the total development budget. In every step, some product feedback must be generated. Extreme Programming (XP) [2] by Beck is based on fixed-duration cycles of two weeks. XP requires additional customer value in every increment.

The school of agile product creation approaches is struggling with the architecting process. The leaders of these communities dislike the "big design up-front." However, running on a treadmill of small increments may cause the loss of the "big picture." Architecting and short cycles, however, are not in conflict. The architecture also has to grow in incremental steps.

1.3.5 MILESTONES AND DECISIONS

Project teams are faced with a limited number of large-impact decisions during the creation process. These decisions in general engage the organization with a commitment somewhere in the future. For example:

Ordering of long-lead items where changes in specification or design might obsolete ordered items. Reordering will cause project delay. Using the initially ordered items might decrease system performance.

Ordering of expensive materials where changes in plan, specification, or design might obsolete the ordered materials.

Product announcement cannot be reversed once the outside world has seen the announcement. Note that announcing a new product often impacts the order intake of existing products. Announcing a new product late might cause competitive risks.

Define a minimal set of *large-impact* decisions.
Define the mandatory and supporting information required for the decision.
Schedule a decision after the appropriate phase transition.
Decide explicitly.
Communicate the decision clearly and widely.

Figure 1.13 How to deal with large-impact decisions.

An explicit decision can be planned as a milestone in the project master plan. Information should be available to facilitate the decision: some of the information is mandatory to safeguard the company; some of the information is only supportive.

In general, the mandatory information should be minimized to prevent a rigid and bureaucratic process causing the company to be unresponsive to the outside world. These decisions can be planned after phase transition when most of the required supportive information will be available in an accessible way. Figure 1.13 shows recommendations of how to deal with large-impact decisions.

1.3.6 ORGANIZATION OF THE PRODUCT CREATION PROCESS

The Product Creation Process requires an organizational framework. The organizational framework of the Product Creation Process is independent of the organizational frameworks of the other processes[2]

Hierarchical decomposition

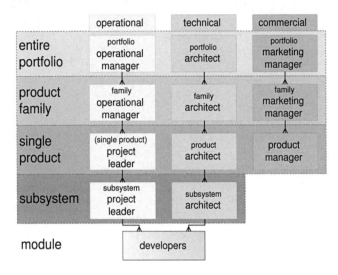

Figure 1.14 The simplified hierarchy of operational entities in the Product Creation Process form the core of this process.

The operational organization is a dominant organizational view of the Product Creation Process. In most organizations, the operations of Product Creation are decomposed into multiple hierarchical levels: at the highest level, the entire product portfolio; at the lowest level, the smallest operational entity, for instance, a subsystem. Note that in Figure 1.14 the hierarchy stops at the subsystem level, although for large developments it can continue into even smaller entities like components

---

[2]Quite often, a strong link is present between the People, Process, and Technology Management Process and the Product Creation Process; using similar frameworks can be quite counterproductive because these processes have quite different aims and characteristics. Most people will operate as part of both organizational frameworks.

or modules. The hierarchy is simply the recursive application of the decomposition principle.

Figure 1.14 is simplified by assuming that a straightforward decomposition can be applied. This assumption is not valid when lower-level entities, for example, subsystems, are used by multiple higher-level entities, for example, systems. For instance, one subsystem that is used in different products breaks this assumption. In Chapter 6, Section 6.1 we elaborate this aspect further.

Further decomposition of the Product Creation Process

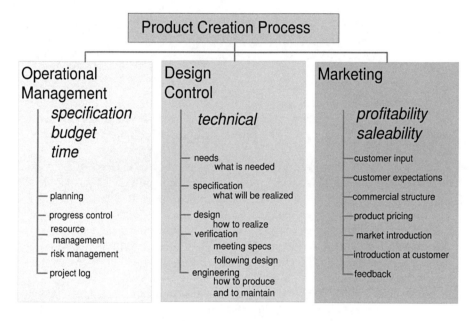

Figure 1.15 Decomposition of the Product Creation Process.

The Product Creation Process can be decomposed into three processes as shown in Figure 1.15:

Marketing: Defining how to obtain a saleable profitable product, starting with listening to customers, followed by managing the customer expectations, introducing the product to the customer, and obtaining customer feedback.

Project Management: Realizing the product in the agreed triangle of
- specification
- resources
- amount of time

Design Control: Specifying and designing the system. The Design Control Process is that part of the Product Creation Process that is close to the conventional R&D activities. It is the content part of the Product Creation Process.

The functions mentioned in Figure 1.14 map directly on the processes in Figure 1.15:

- The *operational* or *project leader* is responsible for *operational management*.
- The *architect* is responsible for *design control*.
- The *marketing* or *product manager* is responsible for *commercial* aspects.

Design Control Process

The ISO 9000 standard has a number of requirements with respect to the *design control* process. The design control process is a core content-oriented process; it is the home base of the systems architect. The systems architect will support project management and the commercial process.

The design control process itself is further decomposed, also shown in Figure 1.15:

Needs The needs express what the stakeholders of the system need, not yet constrained by business or technical considerations. Most development engineers tend to forget the original needs after several iterations of commercial and technical trade-offs.

Specification The specification describes what will be realized in terms of functionality and performance. This specification is the agreement with all stakeholders. The difference between the needs and the specification is that, in the specification, all trade-offs have been made. See also Chapter 3, Section 3.2, where we elaborate on the process of needs analysis and requirements management.

Design The design is the description of how the specification will be realized. For instance, the physical and functional decomposition and the budgets for critical technical resources belong to the design.

Engineering The engineering process provides the foundation upon which the Customer-Oriented Process works for the entire life cycle of the product. The documentation generated in the engineering process is the output of the Product Creation Process.

Verification The verification process verifies that the implementation meets the specification in the way it is specified in the design.

Needs, specification, and design are documented in development documents. The main function of these documents is to streamline the Product Creation Process. During this process, these are living documents fulfilling an important communication function, while at the same time playing an important role in the control aspect of the design process.

Operational Management

Operational management is governed by a simple set of rules; see Figure 1.16. These rules combine a number of very tightly coupled responsibilities in one function to

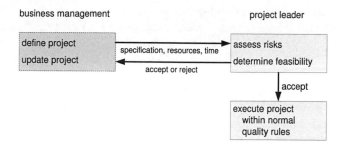

Figure 1.16 Commitment of the operational leader to the project charter.

Figure 1.17 The Operational Triangle of responsibilities; The operational leader commits to the timely delivery of the specification within the agreed budget, with the "standard" quality level.

enable a dynamic balancing act by the operational leader. These responsibilities form the operational triangle as shown in Figure 1.17.

The rules ensure that the operational leader takes ownership of the timely delivery of the specification within the agreed budget, with the "standard" quality level. Transfer of one of these responsibilities to another person changes the system in an open-loop system[3].

Marketing

The marketing manager knows the market: who are potential customers, what are their needs, what is of value in the market, what are commercial partners, what is the competition, etc. This knowledge is "future" oriented and is used to make choices

---

[3]Many conventional development organizations have severe problems with this aspect. A common pitfall is that either the quality responsibility or the resource (budget) responsibility is transferred to the People, Process, and Technology Management Process. The effect is that excuses are present for every deviation of the commitment, for instance, *I missed the timing because the people were working on non-project activities.*

for future products such as what are feasible products, what are the features and performance figures for these products, based on choices where value and cost are in a healthy balance. Hence, the marketing manager is involved in packaging and pricing of products and options. A good marketing manager looks at options broader than the current products. Most innovations are not "more of the same" but are derived from new opportunities, either technical or in the application.

Note that most sales managers are much more backward oriented: we sell what we have to customers who know existing systems. Good salespersons are often not good marketing persons!

Product Creation Teams

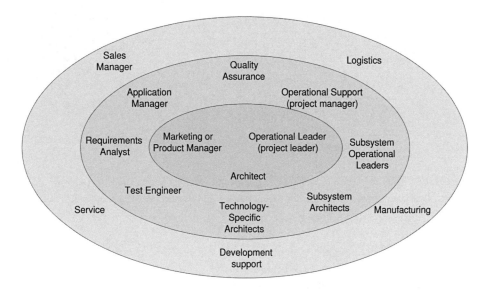

Figure 1.18   The operational teams managing the Product Creation Process.

So far we have discussed *Operational management*, *Design Control*, and *Marketing*. However, in the Product Creation Process, more specialized functions can be present. Figure 1.18 shows a number of more specialized functions as part of a number of concentric operational teams. The amount of specialization depends on the size of the operation. In very small developments, none of the specializations exists, and the roles of project leader and architect are combined in a single person.

1.4   INTERMEZZO: THE IMPORTANCE OF FEEDBACK

1.4.1   INTRODUCTION

Feedback is a universal principle that is applied in highly technical domains such as control engineering, but also in social sciences. This Intermezzo discusses feedback as part of the Systems Architecting Process and explains its importance.

1.4.2   WHY FEEDBACK?

Control

Feedback is used in control systems to ensure that the actual direction corresponds to the desired direction. In general, the deviation from the desired direction grows exponentially in time; see Figure 1.19.

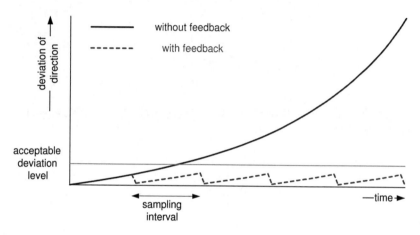

Figure 1.19   The deviation of the actual direction of product development with respect to the desired direction as function of the time.

Many control systems implement a feedback loop to force the system back in the desired direction. Figure 1.19 also shows the effect of a discrete feedback system over time. It will be clear that the sampling interval is determined by the time constant of the deviation and the acceptable deviation level.

Product development can be seen as an ordinary system that can be controlled with a feedback loop like a technical control system. Product development without feedback loops results in products that are out of specification (too late, too slow, too expensive, too heavy, etc). Sound development processes contain (often multiple) feedback loops.

Learning

Human beings learn from their mistakes, provided that they are aware of them. Feedback is the starting point of the learning process, because it provides for the detection *of mistakes. Efficiency of individuals and organizations can be increased by learning. Without learning, similar mistakes are repeated: a waste of resources.*

Applicability

*The principle of feedback can be applied to **any** activity. The higher the uncertainty or the larger the duration of an activity is, the more important feedback becomes.*

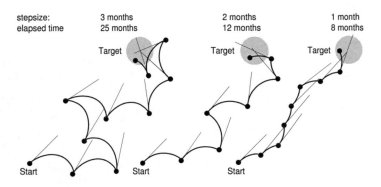

Figure 1.20   Example with different feedback cycles (1, 2, and 3 months) showing the time to market decrease with shorter feedback cycles,

Figure 1.20 shows an example of a development with three different feedback cycle times: months three, two, and one, respectively. The three-month feedback cycle results in a project duration of 25 months. Decreasing the feedback cycle to 2 months brings the total project duration down to 12 months. The one-month feedback cycle gives a total time of only 8 months.

Note that this model is oversimplified and not calibrated. For instance, the degree of deviation from the goal depends on many factors such as the maturity of the development organization, the technology, the product, and the market. This simple model ignores the cost of obtaining feedback, but it clearly illustrates the essence of short feedback cycles.

1.4.3   THEORY VERSUS PRACTICE

Systems architecting is partially a conceptual activity. The concepts stay theoretical as long as they are only part of presentations or specifications. Some architecting schools promote the system architecting function as strategic, providing direction, without being drowned in operational problems. A second school promotes an architect who is active in the definition phase of a product as well as in the verification phase. We argue a third direction: architecting has to be done during the entire development life cycle. In practice, many architects function still in a fourth way: entirely in the technical domain. Figure 1.21 visualizes the four different schools as function of the process phase.

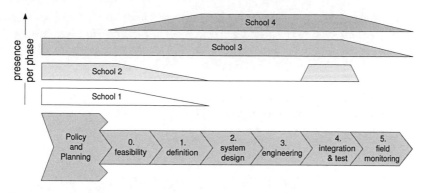

Figure 1.21 The four different schools of architecting, showing the presence of the architect in relation to the policy and planning process and the product creation process.

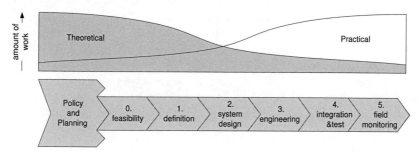

Figure 1.22 Theoretical versus practical system architecture work in relation to the development life cycle.

Figure 1.22 shows the amount of "theoretical" work and the amount of "practical" work also as function of the process phase. We use the term "theoretical" for concepts in presentations or specifications that have not been exposed to the physical world. Similarly, "practical" is used for work where the design has been realized and tested.

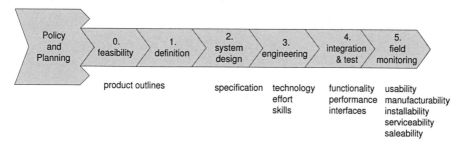

Figure 1.23 Feedback per development phase

A number of feedback loops can be closed during the Product Creation Process.

Normally, the next phase in the process provides feedback to the previous phase in the process. This phase transition feedback is often present. However, feedback from the next phase is a rather indirect measure for the desired direction. The next step provides feedback on the usefulness of the input to continue the work, but user satisfaction and market success cannot be assessed by the next step.

The feedback for the theoretical work comes from the practical work. Figure 1.23 shows the feedback per development phase. This figure makes it immediately clear that the amount of feedback is proportional to the amount of practical work going on.

1.5 THE SYSTEMS ARCHITECTING PROCESS

1.5.1 INTRODUCTION

This section positions the Systems Architecting Process in a wider business scope. This positioning is intended to help understand the process itself and the role of the systems architect (or team of systems architects).

We focus on systems architecting within organizations that create and build systems consisting of hardware and software. Although other product areas such as solution providers, services, courseware, etc. also need systems architects, the process structure will deviate from the structure as presented here. See Section 1.6, Intermezzo: Products, Projects, and Services, for an elaboration of these other architecting models.

1.5.2 SYSTEMS ARCHITECTING IN THE BUSINESS CONTEXT

Figure 1.24 shows the main activities of the Systems Architecting Process as an overlay over business decomposition.

Processes are goal oriented, as discussed in the Intermezzo Section 1.2. The process decomposition is not orthogonal; several processes are overlapping. The Systems Architecting Process is a clear example of such nonorthogonal decomposition. Figure 1.25 shows a map of the Systems Architecting Process and neighboring processes. Many processes such as manufacturing engineering and service engineering have been left out of the map, although these processes also have a high architecting relevance.

Both figures make it clear that the Systems Architecting Process contributes heavily to the Product Creation Process, while it also plays an essential role in the Policy and Planning Process. Both contributions are strongly coupled; see Figure 1.26.

The Systems Architecting Process bridges the gap between the Product Creation Process and the Policy and Planning Process. In many organizations, this link is missing. The absence of this link results in

- reinventing a (different) product positioning during the Product Creation Process, with a limited context view.
- policies that are severely handicapped by a lack of practicality or realism.

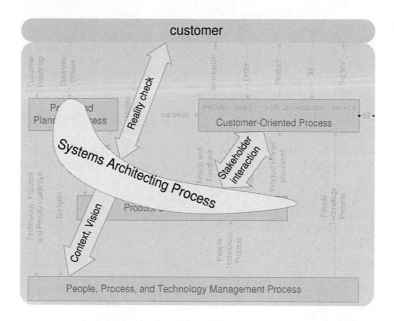

Figure 1.24   The main Systems Architecting activities in the business context.

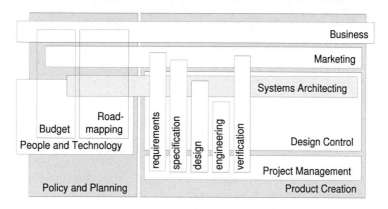

Figure 1.25   Map of the Systems Architecting Process and neighboring processes.

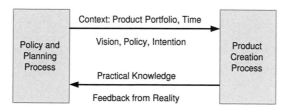

Figure 1.26   Contribution of Systems Architecting to the coupling of the Policy and Planning Process and the Product Creation Process.

The overview created by the Systems Architecting Process also helps in establishing a technology policy.

1.5.3 PURPOSE OF THE SYSTEMS ARCHITECTING PROCESS

Every business larger than a few people enables efficient concurrent work of these people by dividing the tasks into smaller, more specialized jobs: the *decomposition principle* in action. This decomposition of responsibilities requires an opposing force integrating the activities into a useful overall business result. Several integrating processes are active in parallel, such as project management, commercial management, etc.

The Systems Architecting Process is responsible for

- The *integral technical* aspects of the Product Creation Process, from requirement to deployment.
- The *integral technical* vision and synergy in the Policy and Planning Process.

The System Architecting Process is striving for an optimal overall business result by creating and maintaining the key issues such as a balanced and consistent design, selection of the least complex solution, and satisfaction of the stakeholders.

The System Architecting Process is balancing, amongst others,

- External and internal requirements
- Short-term needs and long-term interests
- Efforts and risks from requirements to verification
- Mutual influence of detailed designs
- Value and costs

Such a balance is obtained by making trade-offs, for example, *performance* versus *qualities* versus *functionality*, or *synergy* versus *specific solutions*.

It is the purpose of the Systems Architecting Process to maintain consistency throughout the entire system, from roadmap and requirement to implementation and verification. On top of this consistency, integrity in time must be ensured.

An enabling factor for an optimal result is *simplicity* of all technical aspects. Any unnecessary complexity is a risk for the final result and lowers overall efficiency.

1.5.4 THE SYSTEMS ARCHITECT AS PROCESS OWNER

The owner of the Systems Architecting Process is the Systems Architect or the Systems Architecting Team. Many other people are involved in the System Architecting Process.

The systems architect or the team members spend the majority of their time, about 80%, in the Product Creation Process. From the remaining time, the majority is spent in the Policy and Planning Process. In Section 1.5.2 it is explained that these processes are strongly coupled. This coupling is for a large part implemented by employing the same people in both processes. Systems architects spend a small amount of time in People, Process, and Technology Management.

1.5.5 SYSTEMS ARCHITECTING IN PRODUCT CREATION CONTEXT

The Systems Architecting Process is striving for consistency and balance from requirement to actual product.

The amount of people working in product creation can vary from a few to tens of thousands of people. All people working on the creation of a new product have only knowledge of a (small) subset of the information. Inconsistencies and local optimal solutions pop up all the time, caused by lack of knowledge of the broader context.

The Systems Architecting Process has to prevent this natural degradation of system quality. Systems Architecting acts proactively by clear and sharp requirements, specification, and system design, as well as reactively by following up the feedback from detailed design, implementation and test.

During the Product Creation Process, many specification and design decisions are made. Quite often, these decisions are made within the scope of that moment. Consecutive decisions can be in contradiction with previous decisions. For instance, a decision is taken to add memory to the product to increase performance, while one month later, the amount of memory is decreased to lower the cost. The Systems Architecting Process maintains integrity over time by looking at decisions from a broader perspective.

1.6   INTERMEZZO: PRODUCTS, PROJECTS, SERVICES

1.6.1   INTRODUCTION

We have focused on the product creation of "box-" like products: products that have a clear physical part. After creation of the "box-" like product, the products are sold as boxes by sales. In the twentieth century, this was one of the dominating models in industry. Another business model is project *delivery: customers order a turnkey solution to be delivered by the supplier.*

At the end of the century, several other types of systems and related business models became increasingly important. An increase of interoperating systems has opened a world of services, for example, traffic information for navigation systems. Services are also systems, but these systems tend to be less tangible, while these service systems often include people, processes, and organizations.

Similarly, System of Systems emerge everywhere. We have become dependent on the interoperation of multiple systems, the system of systems.

1.6.2   PRODUCTS AND PROJECTS

Figure 1.27 shows an axis with, on the left-hand side extreme projects, and, on the right-hand side extreme products. We can characterize the extremes as follows

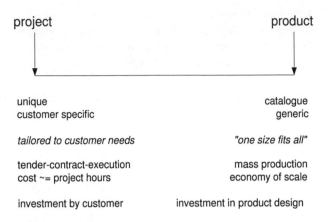

Figure 1.27   Projects versus products.

Projects *are unique to a specific customer. The solution is tailored to the customer's needs. The sales part starts with a tender phase, and the execution phase starts when the contract has been signed. Cost is typically proportional to the number of project hours. In the project business, the customer is the investing party and carries most of the risk.*
Examples of typical projects are buildings, motorways, oil production fields, and cruise ships.

Products *are standardized as part of the sales catalog. Products are designed to be generic, that is to serve multiple customers. The standardization in extreme assumes that "one size fits all." At the same time, standardization enables mass production, while the increased volume of multiple customers provides an economy of scale. Product companies typically invest themselves in new product designs.*
Examples of products are cell phones, televisions, pumps, and MRI scanners.

In practice, business models are less black and white. Figure 1.28 shows a number of forces that lead to convergence between these two extremes. Project organizations see opportunities to increase their margin by harvesting and reusing standardized components or products. For example, in the oil and gas industry, the components that are used in many projects, such as subsea controllers and pumps, are developed as products.

Product organizations adapt their standard products more toward specific customer needs by making them customizable and configurable. Customer support can adapt the product at the customer site to customer-specific needs. For example, MRI scanners are integrated in the hospital workflow by creating hospital-specific clinical protocols.

Figure 1.29 shows a simplified process diagram for project business. The Customer-Oriented Process is replaced by a triplet of processes:

Tender process *prescribes how the specification and price are negotiated with the*

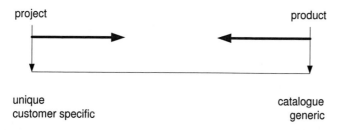

harvest and use
standardized components/products

configuration and customization
customer specific at customer site

project   product

unique   catalogue
customer specific   generic

Figure 1.28   Convergence of projects and products.

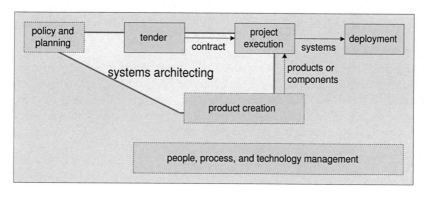

Figure 1.29   Simplified process diagram for project business.

customer.

Execution process *prescribes how the solution is created.*

Deployment process *prescribes how the systems are installed at the customer site and how the operation is started.*

The transition from the tender to the execution phase tends to be a handover, where potentially a lot of knowledge about customer needs is lost. The risk is that feedback from the execution phase cannot be communicated with the original stakeholders from the tender phase.

1.6.3   SERVICES

Figure 1.30 shows an example of a smart phone context. The smart phone as device contains hardware, operating system, and software. The device offers an application infrastructure for many applications that are created by many different parties. The

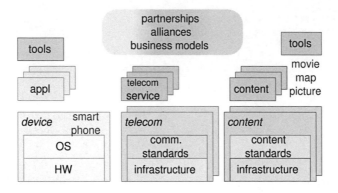

Figure 1.30   Example of extensive complex of services for smart phone type of device.

creation of applications is usually supported by tools.

The applications on the device and telecom services facilitate content services in the broader world, for example, a location service based on position, map, and directory information.

Device builders have to cooperate with the telecom world and the content world to create a saleable device. Developing telecom services and developing content services can also be seen as the creation of systems. However, the world of content creation is much less technical. Forging and nurturing partnerships and alliances is crucial, as well as the development of business models.

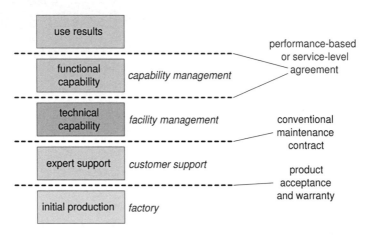

Figure 1.31   Model of operational services showing that the boundary between provider and customer can be defined at different levels.

The type of deliverable and the related business model is also shifting. The conventional model is that the supplier delivers a product according to specification. The relation with the customer stops once the product has been delivered. In many business-to-business segments, the relation is extended by offering maintenance contracts. However, in the conventional model, the customer takes ownership of the system and takes care of maintenance and changes. The bottom two layers in Figure 1.31 represent the conventional business models.

In business-to-business situations, the system that is delivered will be managed by a facilitation or technical department; for example, in hospitals, the radiology equipment is supported by technical hospital staff. The actual operation of the system is done by application experts, in the hospital, for example, the radiology equipment is run by dedicated clinical staff and radiologists. The radiology department provides an imaging and diagnosis capability to the referring physicians.

The equipment manufacturer can shift their support "upward" to offer

Facility management providing a technical working and prepared system.
Capability management where the whole capability, such as diagnostic imaging, is offered.

The consequence of this shift is that the supplier creates a recurring revenue stream. The integral consequence for customer and supplier is that incentives are changing.

For example, when the supplier is responsible for a constant performance, then the supplier might decide to upgrade the equipment much more regularly. The supplier also gets an incentive to minimize downtime and maintenance costs.

The process structure might be adapted to facilitate the service development. Service development, both for the content type as well as for the operational type, require many less technical, more political, social, and economic development activities.

1.6.4 SYSTEM OF SYSTEMS

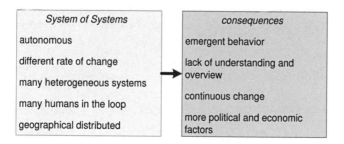

Figure 1.32 System of Systems and the consequences of this approach.

Today's society depends heavily on the interoperability of many systems. We recognize that the solution can be created by interoperability of multiple systems, the so called System of Systems. See Figure 1.32 for the characteristics of System of Systems and the consequences of this approach. The System of Systems can be seen as a supersystem.

Examples of system of systems are

Military capabilities, *where, among others, planes, tanks, guns, officers, soldiers, and sensors are interconnected.*

Health care treatment room, *for example, operating theater or catheterization laboratory, where respiratory and physiology monitors, surgical tools, clinical support systems, nurses, surgeons, etc., collectively perform the treatment function.*

The individual systems in a System of Systems can operate autonomously. Most often these systems have not been created with this specific supersystem in mind. The individual systems follow their own life cycles, with different rates of change. The systems can be quite heterogeneous (large, small, expensive, low cost, reusable, disposable, fragile, robust, etc.). Every system has its own human–machine interface and its own control paradigm. The geographical location of the systems can be distributed and may change.

These characteristics have several consequences. The most dominant consequence is that the supersystem is so complex that nobody has the understanding and the overview of the whole. Hence, nobody can predict what will happen, and we get the so-called emergent behavior. The amount of systems with their different change rates and the amount of humans create a supersystem that is never exactly the same: it changes continuously. In the larger scope of the System of Systems, many non-technical factors play a role, such as economic or political.

EXERCISES

IN CLASSROOM FOR STUDENTS WITH WORKING EXPERIENCE

Make a map of the operational organization for your product group. Use Figure 1.14 as starting point. Note that organizations in practice might look quite different; please show the actual situation, not the theoretical diagram. Identify who is fulfilling the roles in this organization diagram and put their names in the map. Take one horizontal layer of this diagram and annotate that layer with the relations between the people:

- How do they interact?
- Where are they located (same room, floor, building, etc.)?

IN CLASSROOM FOR STUDENTS WITHOUT WORKING EXPERIENCE

1. Discuss your product and market.
2. Propose process and organization for your company: What does your company do in-house, what does your company outsource?
3. Determine staffing for your company.
4. Consolidate the outcome in one organization diagram.

2 Role and Task of the Systems Architect

2.1 THE AWAKENING OF A SYSTEMS ARCHITECT

2.1.1 INTRODUCTION

Systems architects are scarce. This section describes the observed general growth pattern of systems architects. We hope that the analysis of the characteristics of existing systems architects will facilitate the training of new ones. Reference [18] is one of the founding books describing systems architecting and systems architects.

2.1.2 THE DEVELOPMENT OF SYSTEMS ARCHITECTS

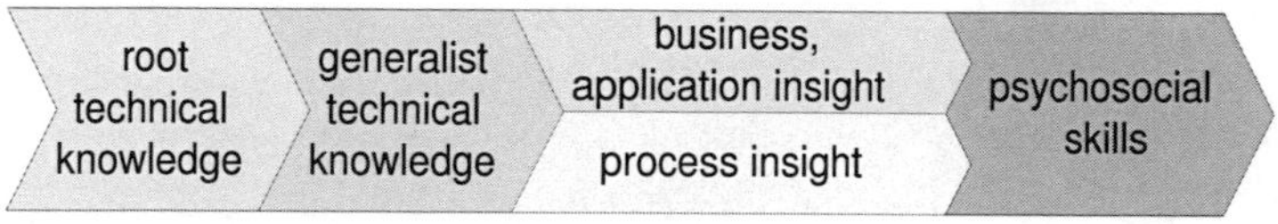

Figure 2.1 Typical development of a systems architect.

Systems architects need a wide range of knowledge, skills, and experience to be effective. Figure 2.1 shows the typical development of a systems architect.

Systems architects are rooted in technology. A thorough understanding of a single technological subject is an essential underpinning. The next step is a broadening of the technical scope. Section 2.1.3 describes the path from a monodisciplinary specialist to a multidisciplinary systems architect with broad technological knowledge.

When the awakening systems architect has reached technological breadth, then it will become obvious that most encountered problems have a root cause outside technology. The systems architect starts to develop along two main parallel streams:

The business side: market, customers, value, competition, logistics, service aspects
The process side: who is doing what and why, necessitated by the amount of involved stakeholders

During this phase, the systems architects will broaden in these two dimensions. They will view these dimensions from a technological perspective. Again, when a sufficient level of understanding is attained, awareness starts to grow that people behave much less rationally than technical designs. The growing awareness of the psychological and social aspects is the next phase of growth.

2.1.3   GENERALIST VERSUS SPECIALIST

Most developers of complex high-tech products are specialists. They need an in-depth understanding of the applicable technology to effectively guide product development. The decomposition of development work is most often optimized to create a work breakdown enabling these specialists to do their work with as much autonomy as possible.

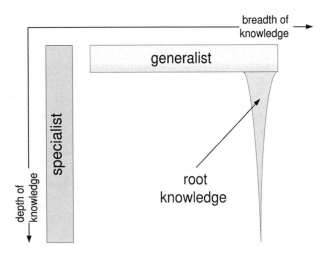

Figure 2.2   Generalist versus specialist; depth versus breadth.

Figure 2.2 is a visualization of the difference between a specialist and a generalist. Most generalists are constrained in the depth of their knowledge by normal human limitations, such as the amount of available time and the finite capacity of the human mind. The figure also shows that a generalist has somewhere roots in detailed technical knowledge. These roots are important for the generalist self since it provides an anchor and a frame of reference. It is also vital in the communication with other specialists because it gives the generalist credibility.

Figure 2.3 shows that both generalists and specialists are needed. Specialists are needed for their in-depth knowledge, while generalists are needed for their general integrating ability. Normally, much more specialists are required than generalists.

There are more functions in the Product Creation Process that benefit from a generalist profile. For instance, the functions of project leader or tester both require a broad area of knowledge.

Architects require a generalist profile since one of their primary functions is to generate top-level specification and design of the system. The step from a specialist to a generalist is not a binary transition. Figure 2.4 shows a more gradual spectrum from specialist to systems architect. The arrows show that intermediate functions exist in larger product developments, forming natural stepping stones for the awakening architect.

Examples of aspect architects are

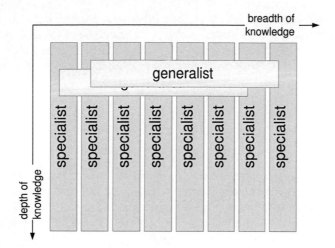

Figure 2.3 Generalists and specialists are both needed in complex products; they have complementary expertise.

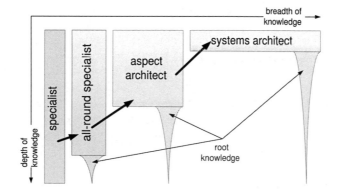

Figure 2.4 Growth in technical breadth; intermediate functions from specialist to systems architect.

Subsystems architects. Subsystems are often the main organizational decomposition. In hardware-intensive systems, subsystems tend to be physical, for example, loader or generator. Typical number of subsystems is between 5 and 15.

Software, mechanics, electronics, or discipline-oriented architects. The architects ensure consistency across physical subsystems.

Function architects take responsibility for one system function, ensuring the soundness of that function.

Quality architects take responsibility for one quality, for example, safety, reliability, or security.

For instance, a software architect needs a significant in-depth knowledge of software engineering and technologies in order to design the software architecture of the entire system. On the other hand, a subsystems architect requires multidisciplinary knowledge. The limited scope of one subsystem reduces the required breadth for the subsystem architect to a hopefully realistic level.

Many products are becoming so complex that a single architect is not capable of covering the entire breadth of the required detailed knowledge areas. In those cases, a team of architects is required, where the architects are complementing each other in knowledge and skills. It is recommended that those architects have complementary roots as well; as this will improve the credibility of the team of architects.

2.2 INTERMEZZO: SYSTEMS TITLES AND ROLES

2.2.1 INTRODUCTION

The following questions are asked frequently during and after the courses:

- *What is the difference between* systems engineers *and* systems architects?
- *Why do all these people have the title* systems architect, *while they actually do not do the work?*

The first questions are also posed in other variants, using titles such as system designer *or* systems manager. *To complicate matters more, there are people who do part of the systems-level work–for example,* requirements analyst, systems analyst, system integrator, *or* system tester– *complementing the systems architect.*

2.2.2 CULTURAL DIFFERENCES IN TERMS

Exactly the same titles are used differently in different companies (or even divisions or product groups within one company), in different domains (for example, defense, automotive, consumer electronics, IT), and geographic regions. No single unified standardized definition is used across companies, domains, and geographies. We do recommend calibration of terminology when entering new territory and to be continuously alert for differences in interpretation even after calibration.

Throughout this book we use the term architecture for the combination of two crucial aspects:

Depth understanding of the system-of-interest, *including product specification, decomposition in subsystems and components, interface management, and function and resource allocation, to create a sound and fitting system that fulfills all qualities (for example, safety, reliability, performance).*

Breadth understanding of the context, *including the customer context and the stakeholders in the value chain, and the life-cycle context from conception to decommissioning and all related business aspects.*

Be aware that the term architect is used often for the system-of-interest *part only. We use the term* system design *for this subset of architecting work; see Section 2.4.*

A major professional society in the systems world is INCOSE, the International Council of Systems Engineering (see www.incose.org*). The systems engineer as depicted by most of INCOSE documents has a very broad function, including the work of the project leader, requirements analyst, systems architect, configuration manager, and quality assurance.*

Another extreme for the definition of systems engineer was used in the medical domain, where this job of the systems engineer was solely the electromechanical design of cables and cabinets.

2.2.3   TITLE VERSUS SKILLS AND ACTUAL JOB

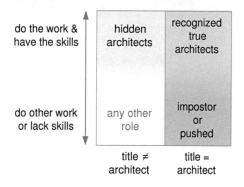

Figure 2.5   Classification of architect related titles.

First of all, we have to distinguish what role or function someone performs and the title that is being used by the people in the context. Figure 2.5 shows a classification obtained by using title as horizontal axis and competence level as vertical axis. Note that the title axis is discrete, while the competence level is continuous. The figure shows four quadrants:

Any other role *(bottom left) for those persons that do not do the job of an architect and do not have the title.*

Impostors or pushed persons *(bottom right) are people lacking the skills or actually not doing the work of an architect but nevertheless have the title architect.*

Note that impostors are those people that actually pursued the title, for example, because of status or income. People that are pushed by their management into this job, when they lack the capability to do it, form another category. People do not become true architects by declaration.

Hidden architects *(top left) can be found in many organizations. They do the work of an architect but are not called architect. These organizations might use different titles or they might not be aware of the systems discipline.*

Recognized true architects *(top right) are those architects that actually do the architecting job skillfully and got the title in recognition.*

2.2.4 SYSTEMS ROLES AND TITLES

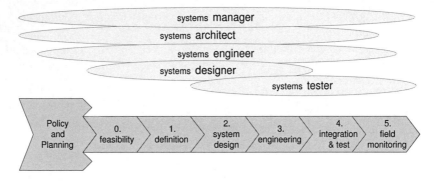

Figure 2.6 System roles mapped on the development life cycle.

In this section we provide a set of roles and relate these to the development life cycles. As explained in the previous sections, these roles can be allocated in different ways, and different terms can be used from the ones shown here. However, the conceptual roles as shown here are quite universal.

Figure 2.6 shows the following roles:

Systems manager *is responsible overall for all systems aspects, ranging from positioning the system strategically in the portfolio and time to final operational performance of the system in the field. Note that such a broad definition does not leave much room for in-depth understanding. An alternate term for this role can be program manager.*

Systems architect, *who combines understanding of the context with in-depth understanding of the solution to create an appropriate system. Note that the architect role combines some perceptive and creative modes of operation with more analytical modes; see Chapter 4, Section 4.2. This mixture limits how deep an architect can go in engineering.*

Systems engineer *is very close to the systems architect, but the emphasis shifts from perceptive and creative more to engineering. With engineering we mean the capability to finalize and document all details required for the later processes such* *as logistics, manufacturing, sales, and customer support. The systems engineer has more the role of* completer finisher, *whereas the architect has more characteristics of the* plant *(see Chapter 10, Section 10.4 for a brief explanation of these Belbin roles). Note also that the systems engineer has limits and will depend on specialized engineers (e.g. mechanical, electrical, or software) to finish the last details of technical product documentation.*

Systems designers *take the product specification as starting point and work on (potential) solutions. Systems designers are "inward" focused, whereas systems architects connect the* outward *and* inward *perspectives.*

System testers *verify that the solution performs as specified. In practice, system testers also need troubleshooting capabilities to diagnose the cause of lacking performance.*

2.3 THE ROLE AND TASK OF THE SYSTEMS ARCHITECT

2.3.1 INTRODUCTION

Architects and organizations are often struggling with the role of the systems architect (or software architect or any other kind of architect). This struggle is partially caused by the intangible nature of the responsibilities of the architect. On the other hand (good) architects are highly appreciated, even if their responsibilities are unclear and their quantifiable output is low.

This section starts with specific deliverables, then discusses the more abstract responsibilities and, finally, discusses the day-to-day activities of an architect.

2.3.2 DELIVERABLES OF THE SYSTEMS ARCHITECT

We start with looking for the tangible output that is expected from architects. Project leaders and program managers do expect deliverables to be finished at appropriate milestones. Most Product Creation Processes define the deliverables of a Systems Architect to be artifacts such as documents or models. These artifacts are symbolized by the stack in Figure 2.7.

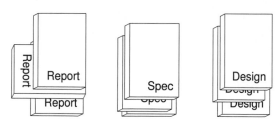

Figure 2.7 Deliverables of Systems Architects consists of artifacts forming a stack of paper when printed.

Figure 2.8 shows the main deliverables of a systems architect more specifically. Quite often, the systems architect does not even produce all deliverables mentioned here, but the architect does take the responsibility for these deliverables by coordinating and integrating the contributions of others. Note that some of these deliverables are part of the Policy and Planning Process.

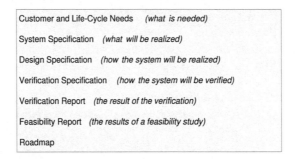

Customer and Life-Cycle Needs   *(what is needed)*

System Specification   *(what will be realized)*

Design Specification   *(how the system will be realized)*

Verification Specification   *(how the system will be verified)*

Verification Report   *(the result of the verification)*

Feasibility Report   *(the results of a feasibility study)*

Roadmap

Figure 2.8   More specific list of deliverables of Systems Architects.

2.3.3   SYSTEMS ARCHITECTS' RESPONSIBILITIES

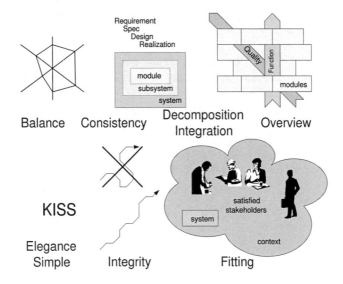

Figure 2.9   The primary responsibilities of systems architects are not tangible, and measurement is difficult.

Systems architects have a limited set of primary responsibilities, as visualized in figure 2.9. The primary responsibilities are:

Balance of system properties as well as internal design properties. The system should be balanced; for example, the cost of subsystems should correspond with its added value in terms of functionality and performance. Architecting is a continuous balancing act in many incomparable dimensions and quantities.

Consistency across many organizational and design boundaries; from needs to implementation details, from system level to detailed implementation.

Decomposition, Integration Decomposition is a default technique to cope with complex and big problems. Decomposing systems in subsystems, subsystems in modules, etc. , is a major responsibility of the architect. In most systems ,many decomposition dimensions are required: physical, logical, functional, and many more; see [13]. The complementary action of decomposition is integration. The integral functioning and performance of the system is the ultimate goal of product creation, which emphasizes the importance of integration. In practice, integration is much more difficult than decomposition; in fact, the architect must decompose in such a way that integration is feasible.

Overview of the entire system and its context helps to make sensible specification and design decisions. The architect should provide overview to all members of the product creation team. Most of these members have a very limited horizon. The architect should facilitate them to make local design decisions by providing proper context information.

Elegance and Simplicity are properties of a "good" architecture. The dangerous aspect of this responsibility is the highly subjective nature of elegance and simplicity. The appreciation of simplicity and elegance should be assessed or acknowledged by others than the architect.

Integrity of the system specification and design over time. The focus of a development team is often wandering over time; sometimes it depends on the hype of the week. The architect is responsible for maintaining a balanced and focused development over time. For instance, when cost price reduction is required, then the architect should keep performance and reliability on the agenda.

Fitting in stakeholder needs and systems context during the entire life cycle is one of the core responsibilities of the architect. The architect must connect depth of knowledge with breadth of knowledge.

We can condense the primary responsibility of the systems architect as ensuring the good functioning of the Systems Architecting Process. In practice, this responsibility is often shared by a team of systems architects, with one chief architect taking the overall responsibility.

The list of primary responsibilities as discussed earlier is suffering from a lack of measurability and is rather intangible. Systems architects also have secondary responsibilities, where these responsibilities are primarily owned by other persons. Most other roles in product creation are more sharply defined, as shown in Figure 2.10. For instance, the business manager is responsible for the business plan and the financial results. The project leader is responsible for the schedule, and hence, for completing the project in time and within budget. The marketing manager is responsible for addressing the relevant markets, and hence, for market share and saleability of the product. The technology manager is responsible for the timely availability of technologies and related tools. The line manager is responsible for the availability of

responsibility	primary owner
business plan, profit	business manager
schedule, resources	project leader
market, salability	marketing manager
technology	technology manager
process, people	line manager
detailed designs	engineers

Figure 2.10 (Incomplete) list of secondary responsibilities of the systems architect and the related primary owner.

the right people with skills and processes to do their job. Finally, the engineers are responsible for the design of their component or module.

2.3.4   WHAT DOES THE SYSTEMS ARCHITECT DO?

Figure 2.11 shows the variety of activities of the day-to-day work of a systems architect. A large amount of time is spent in gathering, filtering, processing, and discussing detailed data in an informal setting. These activities are complemented by more formal activities such as meetings, visits, reviews, etc.

Systems architects are rapidly switching between specific detailed views and abstract higher-level views. The concurrent development of these views is a key characteristic of the way systems architects work.

Abstractions only exist for concrete facts.

Systems architects that stay too long at "high" abstraction levels drift away from reality, by creating their own virtual reality.

Figure 2.12 shows the bottom-up elicitation of higher-level views. Systems architects see a tremendous amount of details, but most of these details are skipped, and a smaller amount is analyzed or discussed. A small subset of these discussed details is shared as an issue with a broader team of designers and architects. Finally, the system architect consolidates the outcome in a limited set of views. The order-of-magnitude numbers cover the activities in one year.

The opposite flow in Figure 2.12 is the implementation of many of the responsibilities of the systems architect. By providing overview, insight, and fact-based direction, a simple, elegant, balanced and consistent design will crystallize, where the integrity of designs goals and solutions are maintained during the project.

A lot of time spent by the architect serves the purpose of communication between many project members. The architect is not only responsible for system integration, but also has an integrating role in the project itself. Architects have to interact a lot with all the people mentioned in Figure 2.10 in order to fulfill their responsibilities.

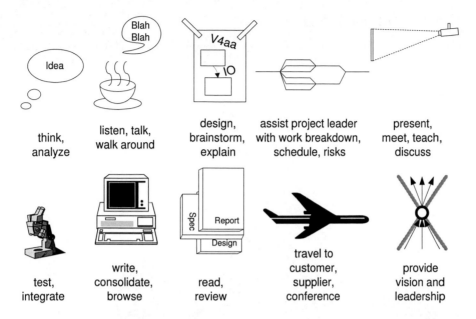

Figure 2.11   The systems architect performs a large amount of activities; where most of the activities are barely visible for the environment, they are crucial to the functioning of architects.

	Quantity per year (order-of-magnitude)	architect time per item
consolidation in deliverables → driving views	10	100 h
meetings → shared issues	10^2	1 h
informal contacts → touched details	10^4	$0.5 - 10$ min
sampling scanning → seen details	$10^5 - 10^6$	$0.1 - 1$ sec
product details	$10^7 - 10^{10}$	
real-world facts	infinite	

Figure 2.12   Bottom-up elicitation of high-level views.

2.3.5 TASK VERSUS ROLE

The task of the systems architect is to generate the agreed upon deliverables; see Section 2.3.2. This measurable output is requested and tracked by the related managers: project leaders and line managers. Many managers judge their architects only by this visible subset of their work.

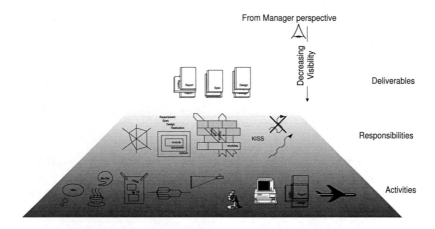

Figure 2.13   The visible outputs versus the (nearly) invisible work at the bottom.

The deliverables are only one of the means to fulfill the systems architect's responsibilities, as described in Section 2.3.3. Systems architects are doing a lot of nearly invisible work to achieve system level goals, their primary responsibility. Figure 2.13 shows this as a pyramid or iceberg: the top is clearly visible; the majority of the work is hidden at the bottom.

2.4 INTERMEZZO: DYNAMIC RANGE OF ABSTRACTION LEVELS IN ARCHITECTING

2.4.1 INTRODUCTION

Systems architects need the capability to "zoom in" and "zoom out." A tremendous dynamic range of abstraction has to be covered from high-level business and customer objectives to detailed design decisions at the engineering level. The system-of-interest itself spans many abstraction levels. However, the architect has to look beyond the system-of-interest itself, toward the customer context, the life cycle, and to related products.

2.4.2 FROM SYSTEM-OF-INTEREST TO CONTEXT

The translation of the product specification of the system-of-interest into detailed monodisciplinary design decisions spans many orders of magnitude. The few statements of performance, cost, and size in the system requirements specification ulti-

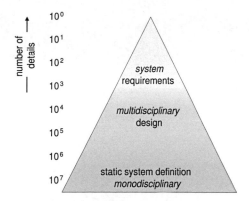

Figure 2.14   Connecting system specification to detailed design.

mately result in millions of details in the technical product description: millions of lines of code, connections, and parts. The technical product description is the accumulation of monodisciplinary formalizations. Figure 2.14 shows this dynamic range as a pyramid with the system at the top and the millions of technical details at the bottom.

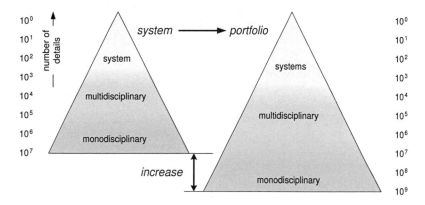

Figure 2.15   From system to product family or portfolio.

The current system-of-interest is most often part of a broader set of products that evolves over time: the product family or portfolio. The aggregate amount of details in the product family or portfolio can be several orders of magnitude larger than the amount of details for one system. Figure 2.15 shows the increase of the dynamic range from system to portfolio.

Architects also have to take the context of the system into account, from both customer as well as business perspective. We can transform the portfolio pyramid from Figure 2.15 into Figure 2.16 to show the number of details of a portfolio in its

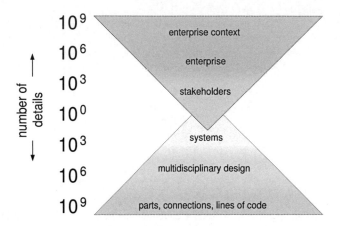

Figure 2.16   Product family in context.

context. The context is also shown as a pyramid, representing the fact that, in the outside world, where systems are actually used, it can be viewed at many levels of abstractions.

2.4.3   ARCHITECTURE AND ARCHITECTING

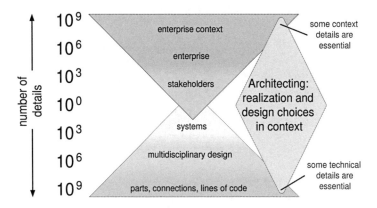

Figure 2.17   Architecture: the essence of system and context.

The challenge of developing an architecture is to capture the essence of both the systems to be built as well as the contexts where systems are being created and used. Figure 2.17 shows that architecture mostly covers the higher abstraction levels. An architecture needs to abstract from most details to facilitate the capture of the essence. Only a simplified description or model can be used at system level to reason *and facilitate communication.*

However, some crucial details either from monodisciplinary area or from the customer or business contexts might have to be included. Quite often the devil is in the detail. Hence, known crucial details are part of an architecture description or model.

Note that architectures do have a scope:

System architecture *captures the essence of a system in its context. Note that the system context includes the product family or portfolio. However, focus of the system architecture is on the system itself and, as such, will position this system in the broader portfolio.*

Family architecture *captures the essence of the family of systems and its context. The focus is now on the family, explaining how different products can support specific market needs and providing guidance to harvest the synergy between products.*

Portfolio architecture *is similar to family but at a higher aggregation level.*

Architecting involves all activities to create an architecture: exploring details in systems and context, communication, design, specification, making decisions, etc. In other words, architecting combines external zoom-in and zoom-out (fact gathering and communication) with internal zoom-in and zoom-out (specification, design, integration).

2.4.4   REVISITING DESIGN AND ENGINEERING

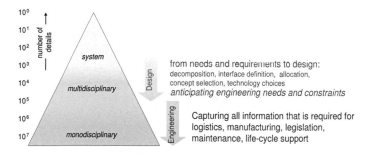

Figure 2.18   Positioning design and engineering in the dynamic range of abstraction levels.

We can revisit the terms design and engineering based on the dynamic range of abstraction levels, as shown in Figure 2.18.

Designing *is the activity to get from needs and requirements for a design: decomposition, interface definition, allocation, concept selection, technology choices, etc. The design has to anticipate engineering needs and constraints.*

Engineering *is capturing all information that is required for the Customer-Oriented Process, such as logistics, manufacturing, legislation, maintenance, life-cycle support.*

Engineering and design mostly takes place internally in the organization, with the exception of the communication with external suppliers.

2.4.5   ARCHITECTING AND DESIGN IN PRACTICE

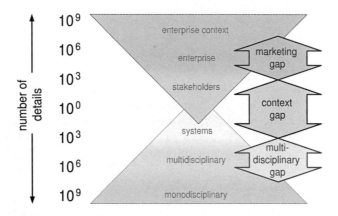

Figure 2.19   Frequently observed gaps in practice.

In practice, several problems can be observed in most organizations that can be explained by "gaps." Figure 2.19 shows some gaps that can be observed in many organizations:

Multidisciplinary gap is the gap between product specification and detailed design decisions.

Context gap is the gap between stakeholders and product specification.

Marketing gap is the gap between the detailed outside world with billions of individuals and our abstracted understanding in terms of stakeholders, concerns, and needs.

Architects have a core role in closing and preventing the multidisciplinary and context gaps. In practice, marketing managers do have the responsibility for the marketing gap with their knowledge of stakeholders, enterprises, and enterprise contexts.

The multidisciplinary gap, from specification to detailed design, is often bridged by experience: older engineers make decisions based on their past experiences. Note that these decisions are often right. The problem is that the implicit nature of these decisions does not facilitate communication, review, or discussion. Worse, this knowledge gradually disappears from the organization, making further evolution even less transparent.

The context gap, how marketing research information relates to choices in product specification, requires an extrovert focus of architects. Early in their careers, many architects look more inward (to design and engineering) and too little outward (to customers and other stakeholders in the Customer-Oriented Process). They take major development steps when they start to address both gaps in a balanced way.

2.5 ARCHITECTING INTERACTION STYLES

2.5.1 INTRODUCTION

A systems architect has to use different interaction styles in different circumstances. In some circumstances, a *leading* style is appropriate, while in other circumstances a *facilitating* style is more effective. Figure 2.20 shows the styles that are discussed in this chapter.

provocation — when in an impasse: provoke / effective when used sparsely

facilitation — especially recommended when new in a field: / contribute to the team, while absorbing new knowledge

leading — provide vision and direction, make choices / risk: followers stop to give the needed feedback

empathic — take the viewpoint of the stakeholder / acknowledge the stakeholder's feelings, needs, concerns

interviewing — investigate by asking questions

whiteboard simulation — invite a few engineers and walk through / the system operation step by step

judo tactics — first listen to the stakeholder and then / explain cost and alternative opportunities

Figure 2.20 Interaction styles for architects.

2.5.2 PROVOCATION

A provocative style can be used by the architect when the discussion is in an impasse. The provocation can be based on taking an extreme viewpoint of one of the stakeholders and confronting the other stakeholders with the consequences. Such a provocation forces the involved stakeholders to formulate their needs more sharply, including the consequences of following the recommendation.

A provocative style should be applied scarcely. Once team members get used to this style, then the style becomes ineffective. Most people do not like to be provoked continuously, so they stop to respond after a few provocations.

2.5.3 FACILITATION

The facilitation style is one where the architect serves the team by facilitating meetings and workshops. Facilitating a meeting means

Preparing the meeting or workshop together with the owner of the meeting: determining the goal, participants, place, agenda, and means

Facilitating the meeting itself: timekeeping, managing the flips, and writing action point and conclusions

Finalizing the meeting: writing a report, creating a presentation of the results, and chasing follow-up actions

The facilitation style is especially useful for architects entering a new domain. The architect provides visible value to the team, while as a spin-off the architect learns a lot about the new domain.

2.5.4 LEADING

A leading style is one where the architect is highly visible. He or she provides vision and direction to the team. The leading architect can be recognized by looking at the followers: if they really follow the architect, then the architect is effective as leader.

The risk of this style is that the team starts to trust the architect decisions too much. Most of the team members have much more knowledge about design issues than the architect does. The architect will often make decisions based on limited knowledge that should be corrected by the specialists with more knowledge. The leading style sometimes inhibits the specialists from opposing the architect. The leading architect must be aware of this effect. Sometimes even invitations to oppose and provocations do not help to loosen up the followers.

2.5.5 EMPATHIC

The empathic style is based on taking the viewpoint of the stakeholder under discussion. This goes much further than the objective rational view. The feelings and emotions of this stakeholder must be taken into account as well. The understanding of the state of mind is communicated back to the stakeholder. The result of this way of interacting is that the architect gets a much better insight into the stakeholders, while at the same time the stakeholders have the feeling that they are taken seriously.

2.5.6 INTERVIEWING

Architects pose many questions; questions are one of the most important instruments of the architect. The interviewing style makes excessive use of questions. The architect uses a priori knowledge to formulate open questions. These open questions must lead to an understanding of the stakeholder's concerns.

The difficult part of this style is to use a priori knowledge in a limited and constructive way. The danger of a priori knowledge is that it limits observation and that suggestive questions are formulated instead of open questions.

2.5.7 WHITEBOARD SIMULATION

The whiteboard simulation style is used in meetings where a few specialists are present. The architect guides the specialists through a use case, where every specialist explains the system behavior from the specialist's viewpoint. For example, the use case can be to push a *next channel* button on the user interface. In this example, the user interface signal will trigger an avalanche of events in the system, going through many layers and propagating into many subsystems.

This guided simulation often reveals a lot of unknown systems behavior, strange dependencies, inefficient sequences, and many more engineering surprises. The normal reactions of the participants are that, after a few steps, they want to redesign the system. The architect should suppress this urge by parking improvements at the side. The main purpose of this style is to build a shared understanding of current design.

2.5.8 JUDO TACTICS

The basis of judo tactics is that the architect starts to listen to the stakeholder, especially when the former feels an urge to contradict the latter. After listening to the stakeholder and acknowledging the validity of the needs, the architect explains the costs and trade-offs. In many cases, the stakeholders have a healthy feeling for value and cost and look for a reasonable balance. Quite often, the result is a decision that the architect wanted to make right at the beginning. However, this style works only if the architect really listens and is willing to take a different direction if needed. It might be that the architect discovers that the value for the stakeholder is much larger than originally assumed!

In many cases, ill communication and bad listening skills block reasonable decisions. The judo style, where the architect starts to listen, avoids this trap.

EXERCISES

Perform a role play in teams of three to four students. In every team we need the roles of *project leader*, *marketing manager*, and *systems architect*. The other team members are observers. In the ideal situation, we have one observer per team.

The teacher provides one actual system. The team has to discuss a very early definition and feasibility. Every team has to create the following deliverables:

Product definition: a very brief summary of the essentials of the specification
Business relevance: a very brief overview why this investment makes sense from the business perspective
Technical feasibility: the core ideas for realization
Initial plan: a very rough plan of when, what, who

Typically, the group will go through the following phases:

1. Use a few minutes to allocate the roles.
2. Spend 5 minutes individually to think about one's own role in relation to the product. For example, the marketing manager will have to think about customer needs, timing, and prices. The architect will explore technologies and potential solutions, and will prepare questions to ask the marketing manager. The project leader will explore required budget, resources, and time.

3. Have a group meeting of about 30 minutes about the deliverables
4. Use 5 to 10 minutes to make a flipchart-based presentation in the classroom
5. The observer also makes one flipchart to explain the observations. Note that it is wise if the team discusses these observations before these are presented plenary.
6. The teams report their deliverables and observations plenary, 5 minutes per team.

3 From Customer Understanding to Requirements

3.1 CAFCR+: A MODEL TO RELATE CUSTOMER NEEDS TO SYSTEM REALIZATION

3.1.1 INTRODUCTION

A simple reference model is used to enable the understanding of the inside of a system in relation to its context. The next section describes the model. We finish the introduction with a few remarks about the naming of the model.

An early tutorial [16] of this model used the concatenation of the first letters of the views in this model to form the acronym "CAFCR" (Customer Objectives, Application, Functional, Conceptual, and Realization). This acronym is used so often within the company that changing the acronym has become impossible. We keep the name constant despite the fact that better names for some of the views have been proposed. The weakest name of the views is *Functional* because this view also contains the so-called *nonfunctional* requirements. A better name for this view is the Black Box view, expressing the fact that the system is described from outside without assumptions about the internals.

The model has been used effectively in a wide variety of applications, ranging from motor way management systems to component models for audio/video streaming. The model is not a silver bullet and should be applied only if it helps the design team.

3.1.2 THE CAFCR MODEL

A useful top-level decomposition of an architecture is provided by the so-called "CAFCR" model, as shown in Figure 3.1. The *Customer Objectives* view and the *Application* view provide the **why** from the customer. The *Functional* view describes the **what** of the product, which includes (despite the name) also the *nonfunctional* requirements. The **how** of the product is described in the *Conceptual* and *Realization* view, where the conceptual view is changing less in time than the fast-changing realization (Moore's law!).

The job of the architect is to integrate these views in a consistent and balanced way. Architects do this job by *frequent viewpoint hopping*: looking at the problem from many different viewpoints, and sampling the problem and solution space in order to build up an understanding of the business: top-down (objective driven, based

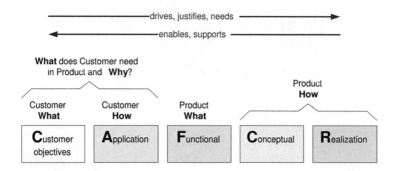

Figure 3.1 The "CAFCR" model.

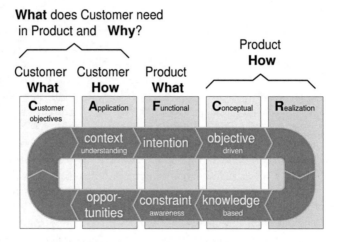

Figure 3.2 Five viewpoints for an architecture. The task of the architect is to integrate all these viewpoints in order to get a *valuable*, *usable*, and *feasible* product.

on intention and context understanding) in combination with bottom-up (constraint aware, identifying opportunities, knowledge based); see Figure 3.2.

In other words, the views must be used concurrently, not top-down like the waterfall model. However, at the end, a consistent story-line must be available, where the justification and the needs are expressed at the customer side, while the technical solution side enables and supports the customer side.

The model will be used to provide a next level of reference models and submethods. Although the five views are presented here as sharp disjointed views, many subsequent models and methods do not fit entirely in one single view. This in itself is not a problem–the model is a means to build up understanding; it is not a goal in itself.

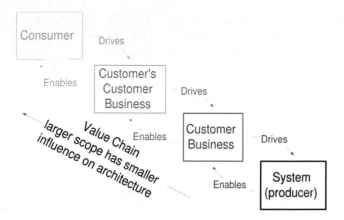

Figure 3.3   CAFCR can be applied recursively.

3.1.3   WHO IS THE CUSTOMER?

The term *customer* is easily used, but it is far from trivial to determine the customer. The position in the value chain shows that multiple customers are involved. In Figure 3.3, multiple customers are addressed by applying the CAFCR model recursively.

Using the term "The customer" is a gross generalization. Marketing managers make a classification of customers by means of market segmentation. The method recommended is to start building up insight in segmentation by making specific choices for a customer, for example, by selecting specific market segments. Making early assumptions about synergy between market segments can handicap the buildup of customer understanding. This kind of assumptions tends to pollute segmentation and inhibit clear and sharp reasoning.

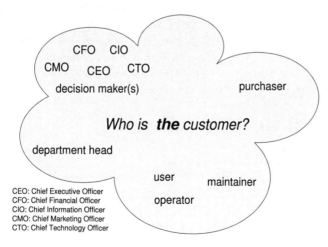

Figure 3.4   Which person is **the** customer?

Many stakeholders are involved for any given customer. Multiple stakeholders are involved even when the customer is a consumer, for example, parents, other family, and friends. Figure 3.4 shows an example of the people involved in a small company. Note that most of these people have different interests with respect to the system.

Market segments are also still tremendous abstractions. Architects have to stay aware all the time of the distance between the abstract models they are using and the reality, with all unique infinitely complex individuals.

3.1.4 LIFE-CYCLE VIEW

The basic CAFCR model relates the customer needs to design decisions. However, in practice we have one more major input for system requirements: life-cycle needs. Figure 3.5 shows the CAFCR+ model that extends the basic CAFCR model with a *Life-Cycle view.*

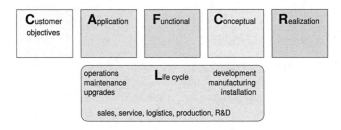

Figure 3.5 CAFCR+ model; Life-Cycle View.

The system life cycle starts with the conception of the system that triggers the development. When the system has been developed, then it can be reproduced by manufacturing, ordered by logistics, installed by service engineers, sold by sales representatives, and supported throughout its lifetime. Once delivered, every produced system goes through a life cycle of its own with scheduled maintenance, unscheduled repairs, upgrades, extensions, and operational support. Many stakeholders are involved in the entire life cycle: sales, service, logistics, production, and research and development. Note that all these stakeholders can be part of the same company or that these functions can be distributed over several companies.

3.2 FUNDAMENTALS OF REQUIREMENTS

3.2.1 INTRODUCTION

The basis of a good systems architecture is the availability and understanding of the needs of all stakeholders. Stakeholder needs are primary inputs for system specification. The terms *requirements elicitation*, *requirements analysis*, and *requirements management* are frequently used as parts of the Product Creation Process that cope with the transformation of needs into specification and design.

3.2.2 DEFINITION OF REQUIREMENTS

The term requirement is quite heavily overloaded in the Product Creation context. *Requirement* is sometimes used as nonobligatory, that is, to express wants or needs. In other cases, it is used as mandatory prescription, for example, a must that will be verified. Obviously, dangerous misunderstandings can grow if some stakeholders interpret a requirement as want, while other stakeholders see it as a must.

We will adopt the following terms to avoid this misunderstanding:

Customer Needs is used for the nonmandatory wishes, wants, and needs.
Product Specification is used for the mandatory characteristics the system must fulfill.

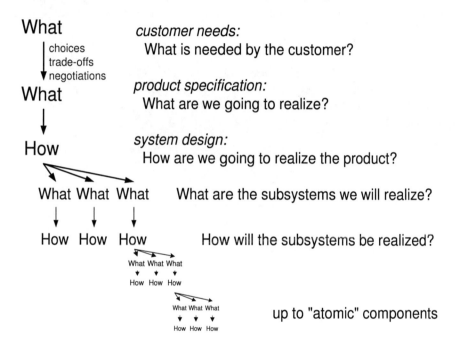

Figure 3.6 The flow of requirements.

In the systems engineering world, the term *Requirements Management* or *Requirements Engineering* is also being used. This term goes beyond the two previous interpretations. The requirements management or engineering process deals with the propagation of the requirements in the product specification toward the requirements of the atomic components. Several propagation steps take place between the product specification and atomic components, such as requirements of the subsystems defined by the first design decomposition. In fact, the requirement definition is recursively applied to every decomposition level similar to product specification and subsystem decomposition.

Figure 3.6 shows the requirements engineering flow. Customer needs are used to determine product specification. Many choices are made going from needs to specification, sometimes by negotiation, sometimes as trade-off. Often the needs are not fully satisfied for mundane reasons such as cost or other constraints. In some cases, product specification exceeds the formulated needs, for instance, anticipating future changes.

Figure 3.6 also shows the separation of specification, *what*, and design, *how*. This separation facilitates clear and sharp decision making, where goals *what* and means *how* are separated. In practice, decisions are often polluted by confusing goals and means.

Another source of requirements is the organization that creates and supplies the product. The needs of the organization itself and of the supply and support chain during the life cycle are described in this chapter as *Life-Cycle Needs*.

3.2.3   SYSTEM AS A BLACK BOX

One of the main characteristics of requirements in product specification is that they describe *what* has to be achieved and not *how* this has to be achieved. In other words, product specification describes the system as *black box*. Figure 3.7 provides a starting point to write a product specification.

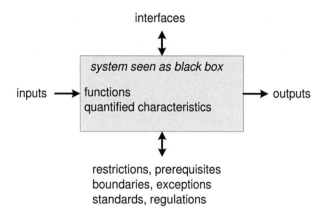

Figure 3.7   System as a black box.

The system is seen as black box– what goes into the box, what comes out, and what functions have to be performed on the inputs to get the outputs. Note that the functions tell something about the black box but without prescribing how to realize them. All interfaces need to be described, interfaces between the system and humans as well as interfaces to other systems. The specification must also quantify desired characteristics, such as how fast, how much, how large, how costly, etc.

Prerequisites and constraints enforced on the system form another class of information in product specification. Further scoping can be done by stating boundaries and desired behavior in case of exceptions. Regulations and standards can be mandatory for a system, in which case compliance with these regulations and standards is part of product specification.

3.2.4 STAKEHOLDERS

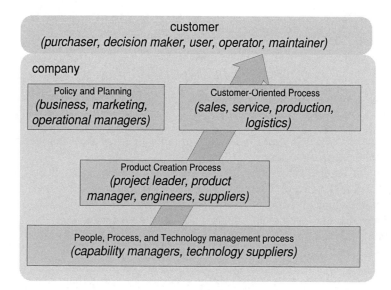

Figure 3.8 A simplified process decomposition of the business. The stakeholders of the requirements are beside the customer self, mainly active in the Customer Oriented Process and the Product Creation Process.

A simplified process model is shown in Figure 3.8. The stakeholders of product specification are, of course, the customers, but also people in the Customer-Oriented Process, the Product Creation Process, the People, Process, and Technology Management Process, and the Policy and Planning Process. The figure gives a number of examples of stakeholders per process.

The customer can be a consumer, but it can also be a business or even a group of businesses. Businesses are complex entities with lots of stakeholders. A good understanding of the customer business is required to identify the customer-stakeholders.

3.2.5 REQUIREMENTS FOR REQUIREMENTS

Standards such as ISO 9000 or methods such as CMM prescribe the requirements for the requirements management process. The left side of Figure 3.9 shows typical requirements for the requirements themselves.

Specific, what is exactly needed? For example, the system shall be *user friendly* is

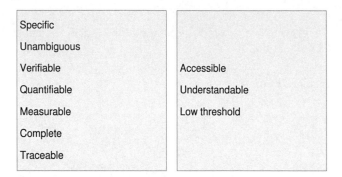

Figure 3.9   Requirements for requirements.

way too generic. Instead, a set of specific requirements is needed that together will contribute to user friendliness.

Unambiguous so that stakeholders do not have different expectations on the outcome. In natural language, statements are quite often context sensitive, making the statement ambiguous.

Verifiable so that the specification can be verified when realized.

Quantifiable is often the way to make requirements verifiable. Quantified requirements also help to make requirements specific.

Measurable to support the verification. Note that not all quantified characteristics can also be measured. For example, in wafer steppers and electron microscopes, many key performance parameters are defined in nanometers or smaller. There are many physical uncertainties to measure such small quantities.

Complete for all main requirements. *Completeness* is a dangerous criterion. In practice, a specification is never complete; it would take infinite time to approach completeness. The real need is that all crucial requirements are specified.

Traceable for all main relations and dependencies. *Traceability* is also a dangerous criterion. Full traceability requires more than infinite time and effort. Understanding how system characteristics contribute to an aggregate performance supports reasoning about changes and decision making.

Unfortunately, these requirements are always biased toward the formal side. A process that fulfills these requirements is, from the theoretical point of view, sound and robust. However, an aspect that is forgotten quite often is that product creation is a human activity, with human capabilities and constraints. The human point of view adds a number of requirements, shown at the right-hand side of Figure 3.9: accessibility, understandability, and a low threshold. These requirements are required for **every** (human) stakeholder.

These requirements, imposed because of the human element, can be conflicting with the requirements prescribed by the management process. Many problems in practice can be traced back to violation of human-imposed requirements. For instance, an abstract description of a customer requirement is formulated such that no real customer can understand this requirement anymore. Lack of understanding is a severe risk because early validation does not take place.

3.3 KEY DRIVER HOW TO

3.3.1 INTRODUCTION

A key-driver graph is a graph that relates the key-drivers (the essential needs) of the customer with the requirements in the product specification. This graph helps to understand the customer better, and the graph helps to assess the importance of requirements. The combination of customer understanding and value assessment makes the graph into an instrument to lead the project.

We will discuss one example, a Motorway management system, and we will discuss a method to create a customer key-driver graph.

3.3.2 EXAMPLE MOTORWAY MANAGEMENT

In this section we discuss an example from practice. The graph discussed here was created in 2000 by a group of marketing managers and systems architects. Creating this version took a few days. Note that we only show and discuss a small part of the entire graph to prevent overload.

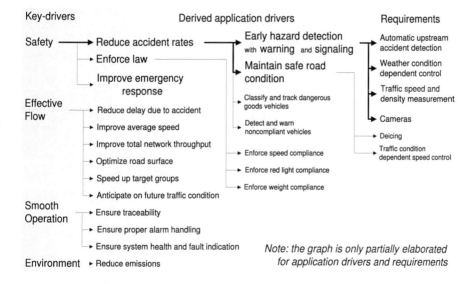

Figure 3.10 The key-driver graph of a motorway management system.

Figure 3.10 shows an example of a key-driver graph of a motorway management system. A motorway management system is one that provides information to traffic controllers, and it allows traffic controllers to take measures on the road or to inform drivers on the road. As drivers, we typically see electronic information and traffic signs that are part of these systems. Also, the cameras along the road are part of such a system.

The key-drivers of a motorway management owner are

Safety for all people on the road: drivers and road maintainers
Effective Flow of the traffic
Smooth Operation of the motorway management
Environmental Protection such as low emissions

To realize these key-drivers, the owner deploys a number of activities. For example, the traffic controllers can improve safety by reducing the accident rate. The accident rate can be reduced by detecting hazards and by warning drivers about the hazards. Examples of hazards are accidents that already have happened and that in turn may trigger new accidents. Other examples of hazards are bad weather conditions or high traffic density. Hence, the automatic detection of accidents and controls that are weather dependent will help to cope with hazards. These functions of the motorway management system will reduce accident rates and improve safety.

Note that the four key-drivers shown here are the key-drivers of the motorway management system. Other systems will also share these concerns but might not have these as key-drivers. For example, smart phones will have a completely different set of key-drivers. Do not use this example as template for your own key-driver graph because it biases the effort.

3.3.3 CAF-VIEWS AND KEY-DRIVERS

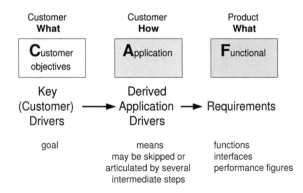

Figure 3.11 The flow from key-drivers via derived application drivers to requirements.

We can capture the essence of the customer world in the *Customer Objectives* view of the CAFCR model by means of customer key-drivers. The customer will organize the way of working such that these key-drivers are achieved. Figure 3.11 shows how the key-drivers as part of the *Customer Objectives* view are supported by application drivers. The application drivers are means to satisfy the customer key-drivers. These application drivers in turn will partially be fulfilled by the system-of-interest. Appropriate requirements, for example, specific functions, interfaces, or performance figures, of the system-of-interest will help the customer use the system to satisfy their customer key-drivers. Key-drivers are thus one of the submethods in the Customer Objectives view.

• Define the scope specific.	in terms of stakeholder or market segments
• Acquire and analyze facts	extract facts from the product specification and ask why questions about the specification of existing products .
• Build a graph of relations between drivers and requirements by means of brainstorming and discussions	where requirements may have multiple drivers
• Obtain feedback	discuss with customers , observe their reactions
• Iterate many times	increased understanding often triggers the move of issues from driver to requirement or vice versa and rephrasing

Figure 3.12 Method to define key-drivers.

Figure 3.12 shows a method to define key-drivers.

Define the scope specific. Identify a specific customer and, within the customer, a specific stakeholder to make the graph. Choosing a customer implies choosing a market segment. A narrow well-defined scope results in a more clear understanding of the customer. The method can be repeated a few times to understand other customers/stakeholders. Products normally have to serve a class of customers. A common pitfall is that the project team too early "averages" the needs; the averaging often compromises the value for specific customers. We recommend first creating some understanding of the target customers before any compromising takes place.

Acquire and analyze facts. We recommend starting to build the graph by looking for known facts. For example, in most organizations, there is already an extensive draft product specification, with many proposed requirements. For every requirement in the draft specification, the *why* question can be asked: "Why does the customer need this feature, what will the customer do with this feature?" Repeating the *why* question relates the requirement in a few steps to a (potential) key-driver. Note that starting with facts often means working bottom-up[1]. When marketing and application managers have a good understanding of the customer, then the facts can also be found in the CA-views, allowing a more top-down approach. Iteration, repeated top-down and bottom-up discussions, is necessary in either case.

Build a graph of relations between drivers and requirements by means of brainstorming and discussions. A great deal of the value of this method is in this

---

[1] Every time that course participants ignore this recommendation and start top-down while lacking customer insight, they come up with a set of too abstract, not usable, key-drivers.

discussion, where team members create a shared understanding of the customer and the product specification. Note that the graph is often many-to-many: one requirement can serve multiple key-drivers, and one key-driver results in many different requirements.

Obtain feedback from customers by showing them the graph and by discussing the graph. Note that it is a good sign when customers dispute the graph, since the graph, in that case, is apparently understandable. When customers say that the graph is OK, then that is often a bad sign, mostly showing that the customer is polite.

Iterate many times top-down and bottom-up. During these iterations, it is quite normal that issues move left to right or opposite due to increased understanding. It is also quite normal that issues are rephrased to sharpen and clarify.

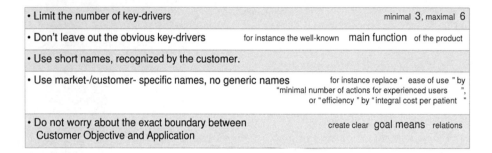

Figure 3.13   Recommendations when defining key-drivers.

Figure 3.13 shows some recommendations with respect to the definition of key-drivers.

Limit the number of key-drivers to maximum six and minimum three. A maximum of six key-drivers is recommended to maintain focus on the essence; the name is on purpose **Key** driver. The minimum (three) avoids oversimplification, and it helps to identify the inherent tensions in the customer world. In real life we always have to balance objectives. For example, we have a strong need to maximize safety and performance, while at the same time we will have cost pressure. A good set of key-drivers also captures the main tensions from the customer perspective.

Do not leave out the obvious key-drivers such as the main function of the product. For example, the communication must be recognizable when discussing smart phones; the focus might be on all kinds of innovative features and services, while the main function is forgotten.

Use short names, recognized by the customer. Key-drivers must be expressed in the language of the customer so that customers recognize and understand them. The risk in teams of engineers is that the terminology drifts away and becomes too abstract or too analytical. Another risk is that descriptions or sentences are used in the graph to explain what is meant. These clarifying texts should not be in the graph itself because the overview function of the graph gets lost. The challenge is to find short labels that resonate with customers.

Use market-/customer- specific names, no generic names. The more specific a name or label, the more it helps in understanding. Generic names facilitate the "escape" of diving into the customer world. For example, the term *ease of use* is way too much of a motherhood statement. Instead, *minimal number of actions (for experienced users)* might be the real issue.

Allocation to Customer Objectives or Application View. Do not worry about the exact boundary between Customer Objective and Application The purpose of the graph is to get a clear separation of goals and means, where goals and means are recursive, an application driver is a means to achieve the customer key-driver, and, at the same time, it is a goal for the functions of the system of interest. Sometimes we need five steps to relate customer key-drivers to requirements, sometimes the relation is obvious and is directly linked. The CAFCR model is a means to think about the architecture; it is not a goal to fit everything right in the different views!

3.4 REQUIREMENTS ELICITATION AND SELECTION

3.4.1 INTRODUCTION

The quality of the system under development depends strongly on the quality of the elicitation process. We can only make a fitting system when we understand the needs of our customer. The outcome of an elicitation process is often an overload of needs. We need a selection process to balance what is needed with all kinds of constraints, such as cost, effort, and time.

3.4.2 VIEWPOINTS ON NEEDS

Needs for a new product can be found in a wide variety of sources. The challenge in identifying needs is, in general, to distinguish a solution for a need from the need itself. Stakeholders, when asked for needs, nearly always answer in terms of a solution. For example, consumers might ask for a *flash-based video recorder*, where the underlying need might be a light-weight, small, portable video recorder. It is the architect's job, together with marketing and product managers, to reconstruct the actual needs from the answers that stakeholders give.

Many complementary viewpoints provide a good collection of needs. Figure 3.14 shows a useful number of viewpoints when collecting needs.

The **key-driver** viewpoint and the **operational** viewpoint are the viewpoints of the stakeholders who are "consuming" or "using" the output of the Product Creation Process. These viewpoints represent the "demand side."

The **roadmap** and the **competition** viewpoints are those that position the products in time and in the market. These viewpoints are important because they emphasize the fact that a product is being created in a dynamic and evolving world. The product context is not static and isolated.

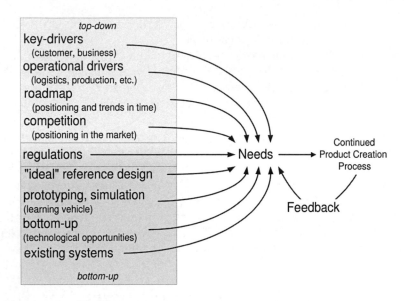

Figure 3.14 Complementary viewpoints to collect needs.

Regulations result in requirements both top-down as well as bottom-up. A top-down example is labor regulations, which can have impact on product functionality and performance. A bottom-up example is materials regulations, which may strongly influence design options, for instance, do not use lead.

The **"ideal" reference design** is the challenge for the architect. What is, in the architect's vision, the perfect solution? From this perfect solution the implicit needs can be reconstructed and added to the collection of needs.

Prototyping or simulations are an important means in communication with customers. This "proactive feedback" is a very effective filter for nice but impractical features on the one hand, and it often uncovers many new requirements. An approach using only concepts easily misses practical constraints and opportunities.

The **bottom-up** viewpoint is one where the technology is taken as the starting point. This viewpoint sometimes triggers new opportunities that have been overlooked by the other viewpoints due to an implicit bias toward today's technology.

The **existing system** is one of the most important sources of needs. In fact, it contains the accumulated wisdom of years of practical application. Especially, a large amount of small but practical needs can be extracted from existing systems.

The product specification is a dynamic entity because the world is dynamic: the users change, the competition changes, the technology changes, the company itself changes. For that reason, the **Continuation of the Product Creation Process** will generate input for the specification as well. In fact, nearly all viewpoints are present and relevant during the entire Product Creation Process.

3.4.3   REQUIREMENTS VALUE AND SELECTION

The collection of customer and operational needs is often larger than can be realized in the first release of a product. A selection step is required to generate a product specification with the customer and operational needs as input. Figure 3.15 shows the selection process as black box with its inputs and outputs.

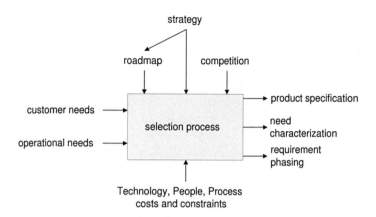

Figure 3.15   The selection process produces a product specification and a phasing and characterization of requirements to prevent repetition of discussion.

The selection process is primarily controlled by the strategy of the company. The strategy determines market, geography, timing, and investments. The roadmap, based on the strategy, is giving context to the selection process for individual products. The reality of the competitive market is the last influencing factor of selection.

Selection will often be constrained by technology, people, and process. The decisions in the selection process require the facts or estimates of these constraints.

During selection, much insight is obtained into needs, the value of needs, and their urgency. We recommend consolidating these insights, for example, by documenting the characterization of needs. The timing insights can be documented in a phased plan for requirements.

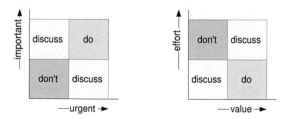

Figure 3.16   Simple methods for a first selection.

The amount of needs can be so large that it is beneficial to quickly filter out the "obvious" requirements. For some needs, it is immediately obvious that they have to be fulfilled anyway, while other needs can be delayed without any problem. Figure 3.16 shows a number of qualitative characterizations of needs, visualized in a two-dimensional matrix. For every quadrant in the matrix, a conclusion is given, a need must be included in the specification, a need has to be discarded, or the need must be discussed further.

This simple qualitative approach can, for instance, be done with the following criteria:

- Importance versus urgency
- Customer value versus effort

In the final selection step, a more detailed analysis step is preferable because this improves the understanding of the needs and results in fewer changes during the development.

A possible way to do this more detailed analysis is to "quantify" the characteristics for every requirement for the most relevant business aspects; see, for examples, Figure 3.17.

• Value for the customer
• (dis)satisfaction level for the customer
• Selling value (How much is the customer willing to pay?)
• Level of differentiation w.r.t. the competition
• Impact on the market share
• Impact on the profit margin
Use relative scale, e.g. 1..5 1=low value, 5 -high value
Ask several knowledgeable people to score
Discussion provides insight   (don't fall in spreadsheet trap)

Figure 3.17   Quantifiable aspects for requirements selection.

These quantifications can be given for the immediate future, but also for the somewhat remote future. In that way insight is obtained into the trend, while this information is also very useful for a discussion of the timing of the different requirements. In [6], a much more elaborate method for requirement evaluation and selection is described.

The output of the requirement characterization and the proposed phasing can be used as input for the next update cycle of the roadmap.

EXERCISES

IN CLASSROOM FOR STUDENTS WITH WORKING EXPERIENCE

Make a key-driver graph for a product you are currently working on. Follow the method and the recommendations. Work in small teams. Start with listing the most important requirements and then work to application drivers and customer key-drivers. The use of yellow note stickers and flipchart pens is recommended: yellow note stickers can easily be moved or removed during the discussion, while flipchart pens force you to limit the label to a few words and numbers. Present the results in one flipchart.

IN CLASSROOM FOR STUDENTS WITHOUT WORKING EXPERIENCE

The teacher will provide a case for this exercise. Use yellow note stickers and flipchart pens during the exercise.

1. Identify the most important requirements of the product. Try to make the requirements as specific as possible; especially quantification helps.
2. Describe what the customer does when using the product. Do this by taking every requirement and answering the *why* question: Why is this requirement needed?
3. Identify the customer key-drivers by repeating the *why* questions.
4. Discuss and improve the entire graph.

THE FINAL RESULT

A good graph can be presented "left-to-right," starting with customer key-drivers and explaining the requirements with a few application stepping stones in between. The graph will, by definition, be far from mature, given the limited time and the classroom setting. In real life, several more iterations with the involvement of external stakeholders will ripen the graph. However, the group now has a much better understanding already of the customer world and, hopefully, is also more aware of their unknowns in that world.

4 Systems Architect Methods and Means

4.1 INTERMEZZO: THE TOOLBOX OF THE SYSTEMS ARCHITECT

4.1.1 INTRODUCTION

The subject of tools for systems architecting creates numerous debates. We will use a broad interpretation of the word tool, including intellectual tools, and low-tech tools (such as pen and paper), but we will also discuss computer-assisted tools. One of the key questions is when to apply what tool. An essential capability for systems architects is to pick an appropriate tool and, if needed, to adapt it to the situation at hand.

We discussed in Sections 1.4 and 2.2 that the role of the systems architect depends on the organizational context. Similarly, there are organizations that force a set of tools on systems architects based on their perceived role and way of working.

We base our discussion of tools on the deliverables, responsibilities, and activities as described in Sections 2.3 and 2.4 (Figures 2.17 and 2.19). Key contribution of systems architects in these sections is the simplification of complicated systems into understandable essentials. The main challenges in achieving this contribution are the heterogeneity of the system and its context, and the uncertainties and unknowns in the system and its context. The goal is to make systems specification and design decisions communicable, and to facilitate debate and reasoning about decisions.

Many organizations move in practice too fast to make extensive use of computer-assisted tools. In consequence, the architects and stakeholders move away from the overview and the understanding of essentials to more detailed concerns (that also have to be addressed!). The purpose of this section is to help understand the impact of tool selection, and especially to bring balance into the application of intellectual tools versus computer-assisted tools.

4.1.2 OVERVIEW OF SYSTEMS ARCHITECTING TOOLS

Figure 4.1 shows an overview and a classification of systems architecting tools. The left-hand side shows the tools that are independent of computers and related software programs. The right-hand side shows tools that depend on computers and specific software. The bottom part of the figure shows some of the standards that impact the selection and application of tools.

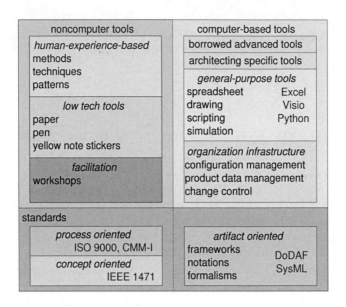

Figure 4.1 Classification of architecting tools.

Human-Experience-Based tools

Experience is crucial to systems architects, who meet new approaches during their entire career, and they build a rich frame of reference by seeing many systems in many circumstances. Reflection on the way of working transforms events into valuable experience: approaches are transformed into methods *and* techniques, *and problems and solutions are transformed into* patterns.

Methods *describe an approach in terms of objective, the order or logic of steps to follow, the techniques that can be applied, and the models, tools, notations, and formalisms that can be supportive.*

Techniques *are ways to address a specific aspect of a problem, for example, how to analyze timing requirements and problems. Techniques can be supported by specific tools and formalisms. A technique may require specific models. For instance, the analysis of response time may require functional flow models.*

Patterns *are recognized problem–solution combinations, including the consideration in what context and circumstances a solution is appropriate. Patterns can be highly technical, for example, the publish–subscribe pattern in software to solve flexibility and extendibility needs. However, patterns can also be high-level organizational or business related, such as considerations about products versus services, as in Figure 1.31 in Chapter 1.*

Low-Tech Tools

Systems architects, like building architects, often make sketches. The sketches are on napkins, paper, flipcharts, whiteboards, using pens, pencils, etc. Sketching is a fast way to express and exchange ideas, and is an essential part of the systems architect's toolbox. Note that, similarly, other low-tech means, such as folded paper, wire frames, and yellow note stickers, provide fast and intuitive ways to express and exchange ideas.

It might be a challenge to capture these sketches for further communication, later reuse, and archiving. However, with today's ubiquitous digital cameras, this is easily captured. Later in the process these sketches get captured electronically in more structured form, for example, in Visio.

Facilitation Tools

Systems architects can contribute to teams by applying facilitation techniques as described in Chapter 2, Section 2.5. An example is the organization of workshops, where teams can explore and share ideas effectively. There are many more facilitation tools and techniques, such as

- *The use of flipcharts to create a common memory on the wall*
- *The use of balanced feedback, such as soliciting benefits and concerns*
- *Working in teams and plenary groups*
- *Preparing meetings together with the leader*
- *Round-robin or random-order contributions to get input from less dominant team members*

Borrowed Advanced Tools

Systems architects cooperate with a large number of experts. Every expert has his or her own set of tools. Sometimes the architect borrows such tool and adapts it to be used at the system level. For example, mechanical engineers are used to tolerance budgets. Systems architects use budgets for many different system qualities (e.g., response time), where granularity of the budget and the algorithms behind the budget have to be adapted to the quality at hand. Most tools in systems architecting find their origin somewhere in another discipline.

Architecting Specific Tools

The problems to be addressed by a tool and the solutions to these problems need to be well defined and repeatable before a computer-assisted tool can be made. The nature of many systems architecting problems is often quite the opposite, with characteristics such as heterogeneous, uncertainties, and unknowns. The systems architecting effort is mostly spent in understanding the problem. Solving well-understood problems in a repeatable and predictable way is the domain of engineering.

Most systems-*specific tools are more engineering related (nailing down all de-tailed information to facilitate the ordering, production, sales, and support of the system) than architecting related. Examples are tools to capture requirements (e.g., Doors), functional and physical architectures such as IDEF0 (e.g., Core), or object-oriented architectures in, for instance, SysML.*

General-Purpose Computer-Based Tools

Architects and engineers use computers all the time for many different purposes. Architects will use a lot of general-purpose tools, such as spreadsheets (e.g., Excel), drawing programs (e.g., Visio), scripting (e.g., Python), or simulation (e.g., Python, MATLAB, or many others).

The general-purpose nature of these tools makes them attractive to architects since that helps them to cope with heterogeneity, unknowns, and uncertainties. The class of more advanced tools can be too restrictive to allow adaptation to the problem at hand and its circumstances.

Tools Prescribed by the Organizational Infrastructure

Organizations do have an engineering tool infrastructure that systems architects can-not ignore. However, systems architects have to decide when and how to interface to the organizational infrastructure. Examples of typical organizational infrastructures are many databases and repositories for engineering-related information:

Configuration management *describing the parts and the rules regarding how the parts can be configured. This repository can be part of a larger system such as an Enterprise Resource Planning (ERP) system (typically, SAP).*
Product Data Management *(PDM) storing all product- and part-related informa-tion required for the Customer-Oriented Process.*
Change Control and Problem Report *databases, where all Change Requests, In-ternal Problem Reports, and Field Problem Reports are stored.*

Systems architects sometimes have to work for some time outside these systems because these systems tend to slow down more creative work full of unknowns and uncertainties. The challenge for project leaders and systems architects is to migrate to these systems at the right moment: using these systems too early slows down work too much, and starting to use them too late might cause loss of information and quality problems.

Process-Oriented Standards

There are many process-oriented standards that influence the way of working of sys-tems architects. For example, the maturity models in CMM-I more or less prescribe most of the tools (configuration management, change control) discussed in previous paragraphs.

Process-oriented standards tend to be agnostic for specific tools. In general, these standards try to capture the best practices from the past in an attempt to prevent past mistakes. *Systems architects in practice suffer when these processes are implemented to the letter rather than the intent. An unintended side effect can be that systems architects are transformed into administrators, while their main contribution is in content rather than administration.*

Concept-Oriented Standards

Some standards try to capture the shared understanding of the architecting discipline. A good example is the IEEE 1471 standard, where the concepts stakeholders, concerns, architecture description, and viewpoints are captured. These standards do not prescribe a way of working but provide a set of concepts and their relations to ease communication.

Artifact-Oriented Standards

In the defense world, several frameworks have been created defining the artifacts that can describe an architecture. Typical examples are the US Department of Defense Architecture Framework (DoDAF) and the British Ministry of Defense Architecture Framework (MoDAF). These frameworks do not define the process, but rather limit themselves to defining the artifacts that may describe the architecture. These standards tend to see the artifacts as electronics artifacts with a significant degree of formalization to facilitate computer assistance.

Part of the Systems Engineering community has transformed UML from the software engineering world into a more systems-oriented modeling language SysML. SysML is a set of formalisms to create artifacts that can be used for computer-assisted tools.

4.1.3   HUMAN VERSUS COMPUTER-ASSISTED TOOLS

One of the main challenges is to decide when and for what to use computer-assisted tools, as stated in the Introduction. Figure 4.2 shows a so-called four quadrant analysis of intellectual (human) tools and computer-assisted tools. The four quadrants are obtained by adding a second dimension: strength and weakness.

Strengths of humans, *based on their intellect, are*
- *To be able to focus on overview.*
- *To be able to identify the essentials.*
- *To understand relationships.*
- *To have insight and intuition.*
- *To be able to synthesize (to combine heterogeneous pieces of information into a meaningful picture).*

Strengths of computers, *based on the current technological level, are*
- *Near-infinite storage capacity.*
- *Near-infinite processing capacity.*
- *The ability to be complete by storing all information.*

	humans	tools	
strength	focus on overview identify essentials understand relationships insight, intuition synthesis	tool dominates focus on details no understanding fragmentation	**weakness**
weakness	limited capacity erroneous behavior incomplete biased	"infinite" storage capacity "infinite" processing capacity complete neutral no errors	**strength**

Figure 4.2 Four quadrant analysis of computerized and human tools.

- *To be neutral, without emotions, opinions, or (political) interests.*
- *To be perfect in execution, making* no errors.

Weaknesses of humans, *inherent to their social and psychological background, are*

- *Storage and processing capacity is limited.*
- *Showing behavior that is erroneous.*
- *Memory is imperfect, information is often* incomplete.
- *Biased for emotional, social or political reasons.*

Weaknesses of computers, *inherent to their mechanistic technical nature, are*

- *The* tool dominates *because there is no "reasonable" flexibility.*
- *The information is in full detail, moving the* focus on details.
- *Computers do not have any* understanding *(garbage in, garbage out).*
- *The data tend to be* fragmented; *only stored relations are present.*

The idea behind the four quadrants is that the weaknesses of humans can be compensated by the strengths of computers and vice versa. If we map these characteristics on the pyramids of Figure 2.17, we then see that the human intellect is required at the higher abstraction levels where we strive for understanding between heterogeneous stakeholders. Computer-assisted tools bring most of their value where large amounts of data have to be managed and processed. Most computer-assisted tools address a limited set of concerns, such that the problem is well defined and the solutions can be applied repeatably and predictably. Many computer-assisted tools are monodiscipline oriented since disciplines capture repeatable knowledge.

4.1.4   FLOW: FROM DATA TO OVERVIEW AND UNDERSTANDING

We have seen in the previous subsection that computer-based tools create most of their value when large amounts of data have to be managed and processed. Other

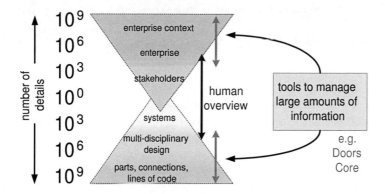

Figure 4.3 Tools Support Processing of Large Amounts of Details.

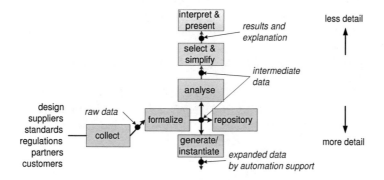

Figure 4.4 From data to understandable information.

discussions that pop up when computer assistance is used is the degree of formalization and the use of automated outputs. Figure 4.4 shows the flow from input data up to the moment that the results are being used by a heterogeneous group of stakeholders. The figure shows the following functions:

Collect data from many inputs, for example, the design, suppliers, standards, regulations, partners, and customers. The output of this function is a collection of raw data— data that still have to be processed to make them useful.

Formalize to be able to enter the data into computer-based tools. The nature of the formalization is to look for appropriate abstractions to capture these data. The consequence of the abstraction is that the amount of detail can decrease slightly, for instance, because repeated data are captured more structurally.

Repositories are used to store the formalized data so that this data can be used for many different purposes. For example, an information model can be stored as entity relationship model plus a data dictionary to capture all formatting details. This information model data can be used to generate data structures and code, it can be used to generate test cases for compliance testing, and the data can be used for analysis.

Generation and Instantiation can be applied on prescriptive data in the repository to generate or instantiate components, stubs, or test harnesses.

Analysis techniques are applied on the data to determine the characteristics of the design. For example, the form, shape, and material characteristics of components can be used to calculate the center of gravity of components and the aggregate of multiple components. Another example is that configurations can be analyzed for feasibility and performance.

Selection and Simplification is a function that is applied by humans (architects or designers) to make the results ripe for communication and discussion. The output of automated analysis techniques is often rather detailed and highly formal, while the essential aspects are hidden in a huge amount of other details.

Interpretation and Presentation are the last steps in making the information accessible and understandable to the broader group of stakeholders. In interpretation, the meaning of the outputs is added, for example: is a center of gravity deviation of 10 mm a problem or is it quite good? The presentation is the format of the output, what visualization will engage the stakeholders, and how to ensure that the information relates to the mental model of the diverse stakeholders.

A common mistake made by engineers is that they show their own intermediate data to stakeholders that use a different mental model themselves. The consequence is that the communication is quite incomplete, and the risk is significant that stakeholders will disconnect or will not give any reaction even when necessary.

Systems architects have to make the last steps of selection, simplification, interpretation, and presentation. Note also that these steps bring their own risks: every simplification is only valid within its limits, so architects are also responsible for monitoring the validity of discussions and decisions in light of the used simplifications.

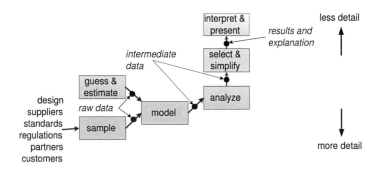

Figure 4.5 Data flow early in the creation process.

Early in the development projects, architects are using a slightly simplified flow to facilitate system specification and design, as shown in Figure 4.5. This figure shows that, early in the process, many estimates and guesses are used, and that less formalization is used. Remember that formalization and computer-based tools are especially relevant when large amounts of data have to be processed and managed. More simple models can be used by architects as long as the amount of information is small.

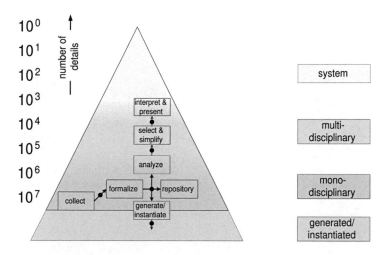

Figure 4.6 Data flow mapped on pyramid.

Figure 4.6 maps the data flow on the pyramid with the abstraction levels. This mapping shows again the relation between the amount of information and the kind of tools to be used: repositories, generator tools, and analysis tools are typically computer assisted, while the intellectual challenges of selection, simplification, interpretation, and presentation are human activities.

Figure 4.7 summarizes these areas of application in the pyramid. The bottom parts of the pyramid with large amount of details can be characterized as more formal and requires more rigor. Formalization requires well-defined problems, data, and operations that are repeatable. The data is machine readable to allow automated tools. The use of repositories facilitates reuse over systems and components.

The upper part of the pyramid is characterized by the combination of quite heterogeneous data with uncertainties and unknowns used by a heterogeneous group of stakeholders with variable backgrounds and concerns. This upper part is less formal and oriented toward communication, discussion, and decision making.

4.2 BASIC WORKING METHODS OF AN ARCHITECT

4.2.1 INTRODUCTION

The basic working methods of architects are covered by a limited set of very generic patterns:

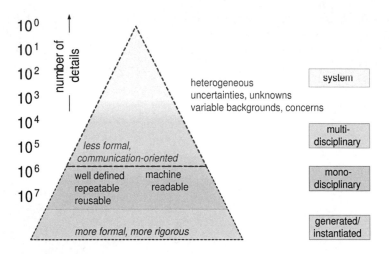

Figure 4.7 Formality levels in pyramid.

Viewpoint hopping: looking at the problem and (potential) solutions from many points of view; see Section 4.2.2.

Decomposition: breaking up a large problem into smaller problems, introducing interfaces and the need for integration; see Section 4.2.3.

Quantification: building up understanding by quantification, from order of magnitude numbers to specifications with acceptable confidence level; see Section 4.2.4.

Decision making: when large parts of data are missing; see Section 4.2.5.

Modeling: as means of communication, documentation, analysis, simulation, decision making, and verification; see Section 4.2.6.

Asking questions: Why, What, How, Who, When, Where, see Section 4.2.7.

Problem solving: An approach to making decisions by understanding, analysis, proposing, deciding, and monitoring; see 4.2.8.

Besides these methods, the architect needs lots of "soft" skills, to be effective with the large amount of different people involved in creating the system. See[18], and Chapters 2 and 10 for additional descriptions of the work and skills of the architect.

4.2.2 VIEWPOINT HOPPING

The architect is looking toward problems and (potential) solutions from many different viewpoints. A small subset of viewpoints is visualized in Figure 4.8, where the viewpoints are shown as stakeholders with their concerns.

The architect is interested in an overall view on the problem, where all these viewpoints are present simultaneously. The limitations of the human brain force the architect to create an overall view by quickly alternating the individual viewpoints. The order in which the viewpoints are alternated is chaotic: problems or opportunities in one viewpoint trigger the switch to a related viewpoint. Figure 4.9 shows a

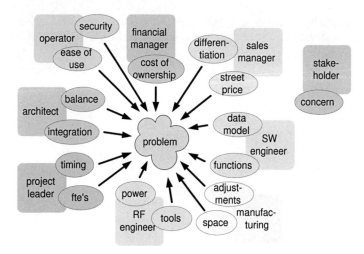

Figure 4.8 A small subset of viewpoints.

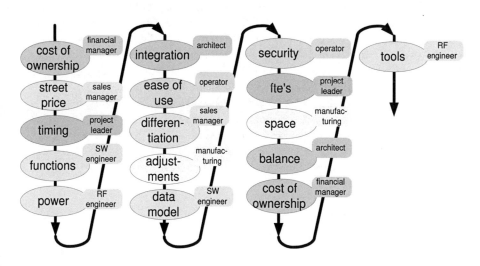

Figure 4.9 Viewpoint hopping.

very short example of viewpoint hopping. This example sequence can take anywhere from minutes to weeks. In a complete product-creation project, the architect makes thousands[1] of these viewpoint changes.

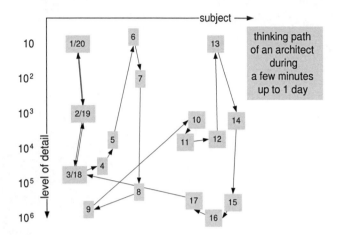

Figure 4.10   The seemingly random exploration path.

Viewpoint hopping is happening quite fast in the head of the architect. Besides changing the viewpoint, the architect is also zooming in and out with respect to the level of detail. The dynamic range of details taken into account is many orders of magnitude. Exploring different subjects and different levels of detail together can be viewed as an exploration path. The exploration path followed by the architect (in the architect's head) appears to be quite random. Figure 4.10 shows an example of an exploration path happening inside the architect's head.

The plane used to show the exploration path has one axis with *subjects*, for example, stakeholders, concerns, functions, qualities, design aspects, etc., while the other axis is *the level of detail*. A very coarse (low level of detail) is, for example, the customer key-driver level. For instance, in component placement machines, the cost per placement of 0.1 millicent/placement is the aggregate result of many detailed design choices. Examples at the very detailed level are lines of code, cycle-accurate simulation data, or component type, material, and size.

Both axes span a tremendous dynamic range, creating a huge space for exploration. Systematic scanning of this space is way too slow. An architect is using two techniques to scan this space that are quite difficult to combine: open perceptive scanning, and scanning while structuring and judging. The open perceptive mode is needed to build understanding and insight. Early structuring and judging is dangerous because it might become a self-fulfilling prophecy. The structuring and judging is required to reach a result in a limited amount of time and effort. See figure 4.11 for these two modes of scanning.

---

[1]Based on observations of other architects and own experience.

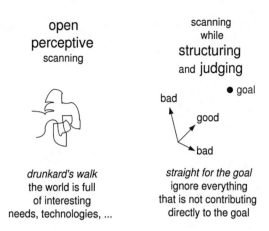

Figure 4.11   Two modes of scanning by an architect.

The scanning approach taken by the architect can be compared with *simulated annealing methods* for optimization [17]. The following is an interesting quote from this book, comparing optimization methods:

Although the analogy is not perfect, there is a sense in which all of the minimization algorithms thus far in this chapter correspond to rapid cooling or quenching. In all cases, we have gone greedily for the quick, nearby solution: From the starting point, go immediately downhill as far as you can go. This, as often remarked above, leads to a local, but not necessarily a global, minimum. Nature's own minimization algorithm is based on a quite different procedure...

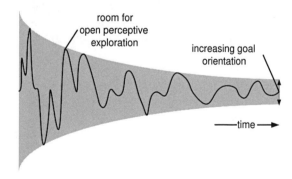

Figure 4.12   Combined open perceptive scanning and goal-oriented scanning.

See also figure 4.12 for the combined scanning path. The perceptive mode is used more early in the project, while at the end of the project the goal-oriented mode is dominant.

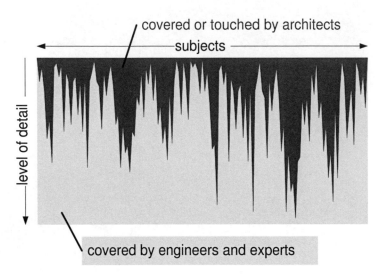

Figure 4.13 The final coverage of the problem and solution space by architect and engineers.

The coverage of the problem and solution space is visualized in figure 4.13. Note that the area covered or touched by the architects is not exclusively covered; engineers will also cover or touch that area partially. Architects need experience to learn when to dig deeper and when to move on to next subjects. Balancing depth and breadth is still largely an art.

4.2.3   DECOMPOSITION AND INTEGRATION

The architect applies a reduction strategy by means of decomposition over and over, as shown in Figure 4.14. Decomposition is a very generic principle. Decomposition can be applied for many different problem and solution dimensions, as will be shown in the later sections.

Whenever something is decomposed, the resulting components will be decoupled by interfaces. The architect will invest time in interfaces since these provide a convenient method to determine system structure and behavior while decoupling the inside of these components from their external behavior.

The true challenge for the architect is to design decompositions that in the end will support an integration of components into a system. Most effort of the architect is concerned with the integrating concepts: how do multiple components work together?

Many stakeholders perceive the decomposition and interface management as the most important contribution of architects. The synthesis or integration part is more difficult and time consuming, and is actually the main contribution of architects.

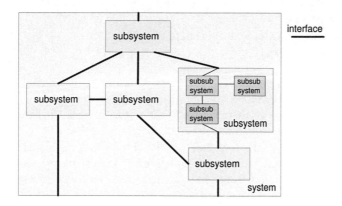

Figure 4.14   Decomposition, interface management, and integration.

4.2.4   QUANTIFICATION

Architects are continuously trying to improve their understanding of problem and solution. This understanding is based on many different interacting insights, such as functionality, behavior, relationships, etc. An important factor in understanding is **quantification**. Quantification helps one to get a grip on the many vague aspects of problem and solution. Many aspects can be quantified, much more than most designers are willing to quantify.

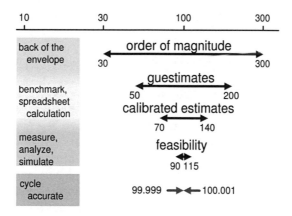

Figure 4.15   Successive quantification refinement.

The accuracy of quantification increases during the project. Figure 4.15 shows the stepwise refinement of quantification. In the first instance, it is important to get a feeling for the problem by quantifying orders of magnitude. For example,

• How large is the targeted customer population?

- What is the amount of money they are willing and able to spend?
- How many pictures/movies do they want to store?
- How much storage and bandwidth is needed?

The order of magnitude numbers can be refined by making back-of-the-envelope calculations, making simple models, and making assumptions and estimates. From this work, it becomes clear where the major uncertainties are and what measurements or other data acquisitions will help to refine the numbers further.

At the bottom of Figure 4.15, the other extreme of the spectrum of quantification is shown: in this example, cycle-accurate simulation of video frame processing results in very accurate numbers. It is a challenge for an architect to bridge these worlds.

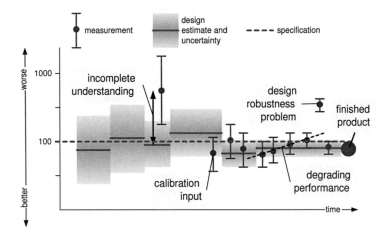

Figure 4.16   Example of the evolution of quantification in time.

Figure 4.16 shows an example of how the quantification evolves in time. The dotted red line represents the required performance as defined in the specification. The shaded area indicates the "paper" value, with its accuracy. The measurements are shown as dots with a range bar. A large difference between paper value and measurement is a clear indication of missing understanding. Later during the implementation, continuous measurements monitor the expected outcome: in this example, a clear degradation is visible. Large jumps in the measurements are an indication of a design that is not robust (small implementation changes cause large performance deviations).

Figure 4.17 shows a graphical example of an "overlay" budget for a wafer stepper. Overlay is the positioning accuracy of the exposed pattern on the wafer. This figure is taken from the *System Design Specification* of the ASML TwinScan system, although for confidentiality reasons some minor modifications have been applied. This budget is based on a model of the overlay functionality in the wafer stepper. The budget is used to provide requirements for subsystems and components. The actual contributions to the overlay are measured during the design and integration process,

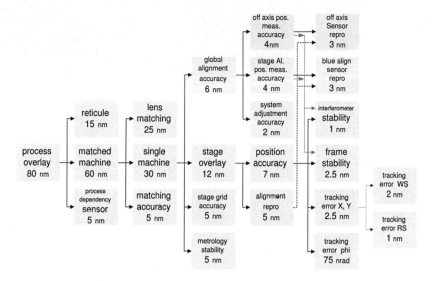

Figure 4.17   Example of a quantified understanding of overlay in a wafer stepper.

on functional models or prototypes. These measurements provide early feedback of the overlay design. If needed, the budget or the design is changed on the basis of this feedback.

4.2.5 COPING WITH UNCERTAINTY

The architect has to make decisions all the time, while most substantiating data is still missing. On top of that, some of the available data will be false, inconsistent, or interpreted wrong.

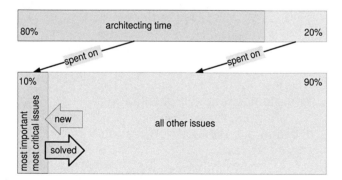

Figure 4.18   The architect focuses on important and critical issues while monitoring other issues.

An important means in making decisions is building up insight, understanding, and overview by means of structuring the problems. The understanding is used to determine important (for product use) and critical (with respect to technical design and implementation) issues. The architect will pay most attention to these *important* and *critical* issues. The other issues are monitored because sometimes minor details turn out to be important or critical issues. Figure 4.18 visualizes the time distribution of the architect: 80% of the time is spent on 10% of the issues.

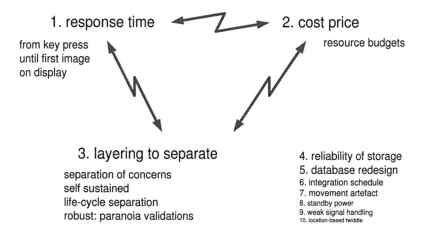

1. response time

from key press
until first image
on display

2. cost price

resource budgets

3. layering to separate

separation of concerns
self sustained
life-cycle separation
robust: paranoia validations

4. reliability of storage
5. database redesign
6. integration schedule
7. movement artefact
8. standby power
9. weak signal handling
10. location-based twiddle

Figure 4.19   Example worry list of an architect.

The architect will, often implicitly, work on the basis of a top 10 issue list, the 10 most relevant (important, urgent, critical) issues. Figure 4.19 shows an example of such a "worry"list.

4.2.6 MODELING

Modeling is one of the most fundamental tools of an architect.

A **model** is
a **simplified** representation of
part of the **real world** used for:

communication, documentation
analysis, simulation,
decision making, and verification

In summary, models are used to obtain insight and understanding, facilitating communication, documentation, analysis, simulation, decision making, and verification. At the same time, the architect is always aware of the (over)simplification applied in every model. A model is very valuable, but every model has its limitations, imposed by simplifications.

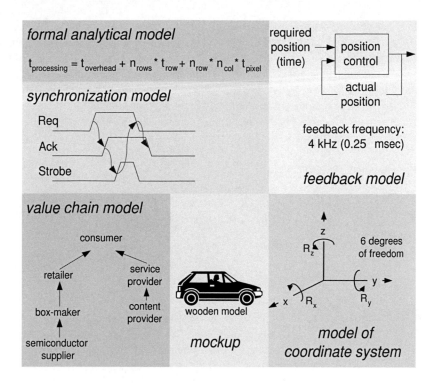

Figure 4.20   Some examples of models.

Models exist in a very rich variety; an arbitrary small selection of models is shown in Figure 4.20.

Models have many different manifestations. Figure 4.21 shows some of the different types of models, expressed in a number of adjectives.

mathematical	
	visual
linguistic	
formal	informal
quantitative	qualitative
detailed	global
concrete	abstract
accurate	approximate
executable	read only
←— rational ——	—— intuitive —→

Figure 4.21   Types of models.

Models can be *mathematical*, expressed in formulas, they can be *linguistic*, expressed in words, or they can be *visual*, captured in diagrams. A model can be formal, where notations, operations, and terms are precisely defined, or informal using plain English and sketches. Quantitative models use meaningful numbers, allowing verification and judgments. Qualitative models show relations and behavior, providing understanding. Concrete models use tangible objects and parameters, while abstract models express mental concepts. Some models can be executed (as a simulation), while other models only make sense for humans reading the model.

4.2.7 WWHWWW QUESTIONS

Why	Who
What	When
How	Where

Figure 4.22 The starting words for questions by the architect.

All "W" questions are an important tool for the architect. Figure 4.22 shows the useful starting words for questions to be asked by an architect.

Why, **what** and **how**, are used over and over in architecting. Why, what, and how are used to determine objectives, rationale, and design. This works highly recursively; a design has objectives and a rationale, and results in smaller designs that again have objectives and rationales. Figure 4.23 shows that the recursion with **why** questions broadens the scope, and recursion with **how** questions opens more details in a smaller scope.

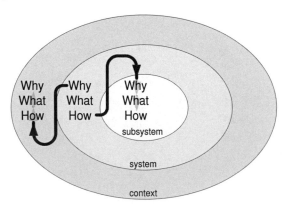

Figure 4.23 Why broadens scope, How opens details.

Who, **where**, and **when** are used somewhat less frequently. Who, where, and when can be used to build up understanding of the context, and are used in cooperation with the project leader to prepare the project plan.

4.2.8   DECISION-MAKING APPROACH

Many specification and design decisions have to be taken during the Product Creation Process. For example, functionality and performance requirements need to be defined, and the way to realize them has to be chosen. Many of these decisions are interrelated and have to be taken at a time when many uncertainties still exist.

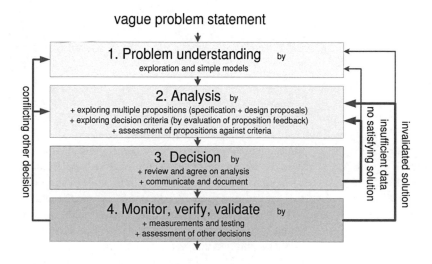

Figure 4.24   Flow from problem to solution.

An approach to making these decisions is the flow depicted in Figure 4.24. The decision process is modeled in four steps. An understanding of the problem is created by the first step ‚*problem understanding*, by exploration of the problem and solution space. Simple models, in the problem space as well as in the solution space, help to create this understanding. The next step is to perform a somewhat more systematic *analysis*. The analysis is often based on *exploring multiple propositions*. The third step is the *decision* itself. The analysis results are reviewed, and the decision is documented and communicated. The last step is to *monitor, verify, and validate* the decision.

The *analysis* involves multiple substeps: *exploring multiple propositions*, *exploring decision criteria*, and *assessing the propositions against the criteria*. Propositions describe both specification (**what**) and design (*how*). Figure 4.25 shows an example of multiple propositions. In this example, a high-performance but high-cost alternative is put beside two lower-performing alternatives. Most criteria get articulated in the discussions about the propositions: "I think that we should choose proposition 2 because ...". The *because* can be reconstructed into a criterion.

throughput	20 p/m	high-performance sensor	350 ns
cost	5 k$	high-speed moves	9 m/s
safety		additional pipelining	
	low cost and performance 1		

throughput	20 p/m	high-performance sensor	300 ns
cost	5 k$	high-speed moves	10 m/s
safety			
	low cost and performance 2		

throughput	25 p/m	highperformance sensor	200 ns
cost	7 k$	high-speed moves	12 m/s
safety		additional collision detector	
	high cost and performance		

Figure 4.25   Multiple propositions.

The decision to choose a proposition is taken on the basis of the analysis results. A review of the analysis results ensures that these results are agreed upon. The decision itself is documented and communicated[2]. In the case of insufficient data or in the absence of a satisfying solution, we have to backtrack to the *analysis* step. Sometimes it is better to revisit the problem statement by going back to the *understanding* step.

Taking a decision requires a lot of follow-up. The decision is in practice based on partial and uncertain data and many assumptions. A significant amount of work is to monitor the consequences and implementation of the decision. Monitoring is partially a *soft skill*, such as actively listening to engineers and, partially, an *engineering activity*, such as measuring and testing. The consequence of a measurement can be that the problem has to be revisited, starting again with the understanding for serious mismatches ("apparently we don't understand the problem at all") or direct into the analysis for smaller mismatches.

The implementation of taken decisions can be disturbed by later decisions. This problem is partially tackled by requirements traceability, where known interdependencies are managed explicitly. In the complex real world, the amount of dependencies is almost infinite. The explicit dependability specifications are inherently incomplete and only partially understood. To cope with the inherent uncertainty about dependencies, an open mind is needed when screening later decisions. A conflict caused by a later decision triggers a revisit of the original problem.

The same flow of activities is used recursively at different levels of detail, as shown in Figure 4.26. A *system* problem will result in a system design, where many design aspects need the same flow of problem-solving activities for the subsystems. This process is repeated for smaller scopes until termination at problems that can be solved directly by an implementation team. The smallest scope of termination is

---

[2]This sounds absolutely trivial, but unfortunately, this step is performed quite poorly in practice.

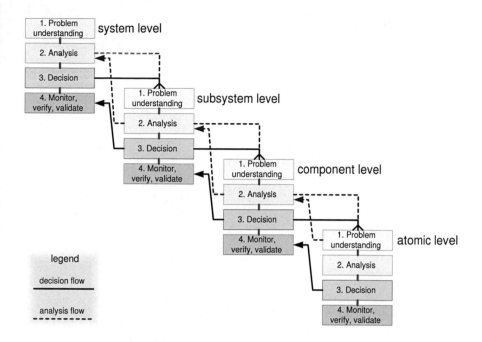

Figure 4.26   Recursive and concurrent application of flow.

denoted as *atomic* level in the figure. Note that more detailed problem solving might have impact on the more global decisions.

4.3   STORY HOW TO

4.3.1   INTRODUCTION

Starting a new product definition often derails in long discussions about generic specification and design issues. Due to lack of reality check, these discussions are very risky, and often way too theoretical. Storytelling followed by specific analysis and design work is a complementary method to do *in-depth* exploration of *parts* of the specification and design.

The method provided here, based on storytelling, is a powerful means to get the product definition quickly into a concrete factual discussion. The method is especially good in improving the communication between the different stakeholders. This communication is tuned to the stakeholders involved in the different CAFCR views: the *story* and *use case* can be exchanged in ways that are understandable for both marketing-oriented people as well as for designers.

Figure 4.27 positions the story in the Customer Objectives View and Application View. A good story combines a clear market vision with a priori realization knowledge. The story itself must be expressed entirely in customer terms; no solution jargon is allowed.

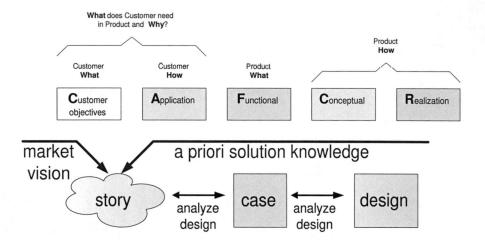

Figure 4.27   From story to design.

4.3.2   HOW TO CREATE A STORY

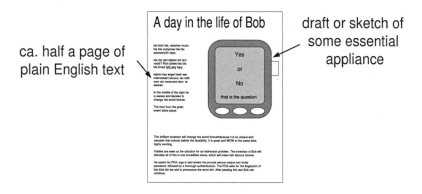

Figure 4.28   Example story layout.

A story is a short, single-page story, as shown in Figure 4.28, preferably illustrated with sketches of the most relevant elements of the story, for instance, the look and feel of the system being used. Other media such as cartoons, animations, video, or demonstrations using mockups can also be used. The *duration* or the *size* of the "story" must be limited to enable focus on the essentials.

Every story has a *purpose*, something the design team wants to learn or explore. The purpose of the story is often in the Conceptual and Realization Views. The *scope* of the story must be chosen carefully. A wide scope is useful to understand a wide context but leaves many details unexplored. An approach is to use recursively refined stories: an overall story setting the context, and a few other stories zooming in on aspects of the overall story.

The story can be written from several *stakeholder viewpoints*. The viewpoints should be carefully chosen. Note that the story is also an important means of communication with customers, marketing managers, and other domain experts. Some of the stakeholder viewpoints are especially useful in this communication.

The *size* of the story is rather critical. Only short stories serve the purpose of discussion catalyst. At the same time, all stakeholders have plenty of questions that can be answered by extending the story. It is recommended that the size of the story be really limited. One way of doing this is by consolidating additional information in a separate document. For instance, in such a document, the point of the story in customer perspective, the purpose of the story in technology exploration, and the implicit assumptions about the customer and system context can be documented.

4.3.3   HOW TO USE A STORY

The story itself must be very accessible to all stakeholders. It must be attractive and appealing to facilitate communication and discussion between those stakeholders. The story is also used as input for a more systematic analysis of the product specification in the Functional View. All functions, performance figures, and quality attributes are extracted from the story into a use case. Note that we use the term *use case* here broader than today's practice. Today's practice limits use cases to a few functions. We recommend to extend a use case into a description of how the system is used *in its context* with a *combination* of functions and a *quantitative description of performance and other qualities*. The analysis results are used to explore the design options.

Normally, several iterations will take place between story, use case, and design exploration. During the first iteration, many questions will be raised in the case analysis and design, which are caused by the story being insufficiently specific. This needs to be addressed by making it more explicit. Care should be taken that the story stays in the Customer and Application Views and that it is not extended too much. The story should be sharpened, in other words, made more explicit, to answer the questions.

After a few iterations, a clear integral overview and understanding emerges for this very specific story. This insight is used as a starting point to create a more complete specification and design.

4.3.4   CRITERIA

Figure 4.29 shows the criteria for a good story. It is recommended that a story be assessed against this checklist and either improved such that it meets all the criteria, or rejected. Fulfillment of these criteria helps to obtain a useful story. The set of five criteria is a necessary but not sufficient set. The value of a story can only be measured in retrospect by determining the contribution of the story to the specification and design process.

Accessible by, understandable The main function of a story is to make the oppor-

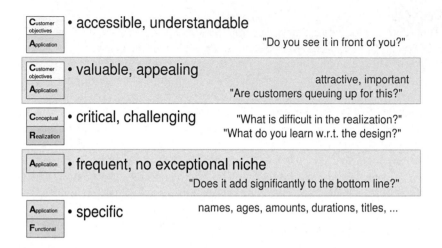

Figure 4.29 Criteria for a good story.

tunity or problem communicable with all the stakeholders. This means that the story must be accessible and understandable for all stakeholders. The description or presentation should be such that all stakeholders can *live through*, *experience*, or *imagine* the story. A "good" story is not a sheet of paper; it is a living story.

Important, valuable, appealing, attractive The opportunity or problem (idea, product, function, or feature) must be significant for the target customers. This means that it should be important for them, or valuable; it should be appealing and attractive.

Most stories fail on this criterion. Some so-so opportunity (whistle and bell-type) is used, where nobody gets really enthusiastic. If this is the case, more creativity is required to change the story to a useful level of importance.

One way to highlight the value of an idea is to tell a story "before" and "after" the introduction of the idea. The contrast between before and after can clarify the value of the idea.

Critical, challenging The purpose of the story is to learn, define, and analyze new products or features. If the implementation of a story is trivial, nothing will be learned. If all other criteria are met and no product exists yet, than just do it because it is clearly a quick win!

If the implementation is challenging, then the story is a good vehicle to study the trade-offs and choices to be made.

Frequent, no exceptional niche Especially in the early exploration, it is important to focus on the main line, the *typical* case. Later in the system design, more specialized cases will be needed to analyze, for instance, more exceptional worst-case situations.

A *typical* case is characterized by being frequent; it should not be an exceptional niche.

Specific The value of a story is the specificity. Most system descriptions are very generic and therefore very powerful, but, at the same time, very nonspecific. A good story provides focus on a single story, one occasion only. In other words, the thread of the story should be very specific.

Specificity can be achieved in social, cultural, emotional, or demographic details, such as names, ages, and locations. "Eleven-year-old Jane in Shanghai" is a very different setting than "Eighty-two-year-old John in an Amsterdam care center." Note that these social, cultural, emotional, or demographic details also help in the engagement of the audience. More analytical stories can be too "sterile" for the audience.

Another form of specificity is information that helps to quantify. For example, using "Doctor Zhivago" as movie content sets the duration to 200 minutes. Stories often need lots of these kinds of detail to facilitate later specification and design analysis. When during the use of the story more quantification is needed, then the story can be modified such that it provides that information.

A good story is in **all** aspects as specific as possible, which means that

- Persons playing a role in the story preferably have a name, age, and other relevant attributes.
- The time and location are specific (if relevant).
- The content is specific (for instance, listening for **2 hours** to songs of **the Beatles**).

There are also a number of pitfalls when writing stories. For example,

Too many issues addressed in a single story. Story writers sometimes want to show multiple possibilities and describe somewhere an escaping paragraph to fit in all the potential goodies (Aardvark works, sleeps, eats, swims, etc., while listening to his Wow56). Simply leave out such a paragraph; it only degrades the focus and value of the story.

Too much sales oriented, obfuscating the later learning issues. A good story should clearly show the benefits of the idea as indicated by the *value* criterion. However, a story must also be realistic and contain sufficient detail for the specification and design analysis later, the *learning* criterion.

Too funny A good story can be humorous. Especially in the classroom setting, students will have fun with creating a story with some humor. However, when stories start to describe grannies of 83 years jumping out of an airplane, then the humor dominates over the story goal.

All these examples typically distract the audience from the intended purpose of the story.

4.3.5 EXAMPLE STORY

Figure 4.30 shows an example of a story for hearing aids. The story first discusses the problem an elderly lady suffers from due to imperfect hearing aids. It continues with postulated new devices that helps her to participate again in an active social life.

Betty is a 70-year-old woman who lives in Eindhoven. Three years ago her husband passed away, and since then, she lives in a home for the elderly. Her two children, Angela and Robert, come and visit her every weekend, often with Betty's grandchildren Ashley and Christopher. As with so many women of her age, Betty is reluctant to touch anything that has a technical appearance. She knows how to operate her television, but a VCR or even a DVD player is way to complex.

When Betty turned 60, she stopped working in a sewing studio. Her work in this noisy environment made her hard-of-hearing with a hearing-loss of 70dB around 2kHz. The rest of the frequency spectrum shows a loss of about 45dB. This is why she had problems understanding her grandchildren and why her children urged her to apply for hearing aids two years ago. Her technophobia (and her first hints or arthritis) inhibit her from changing her hearing aids' batteries. Fortunately, her children can do this every weekend.

This Wednesday, Betty visits the weekly Bingo afternoon in the meeting place of the old-folk's home. It's summer now and the tables are outside. With all those people there, it's a lot of chatter and babble. Two years ago, Betty would never go to the bingo: "I cannot hear a thing when everyone babbles and clatters with the coffee cups. How can I hear the winning numbers?!". Now that she has her new digital hearing instruments, even in the bingo cacophony, she can understand everyone she looks at. Her social life has improved a lot, and she even won the bingo a few times.

source: Roland Mathijssen
Embedded Systems Institute
Eindhoven

That same night, together with her friend Janet, she attends Mozart's opera The Magic Flute. Two years earlier, this would have been one big low rumbly mess, but now she even hears the sparkling high piccolos. Her other friend Carol never joins their visits to the theaters. Carol also has hearing aids; however, hers only "work well" in normal conversations. "When I hear music, it's as if a butcher's knife cuts through my head. It's way too sharp!". So Carol prefers to take her hearing aids out, missing most of the fun. Betty is so happy that her hearing instruments simply know where they are and adapt to their environment.

Figure 4.30   Example of a story.

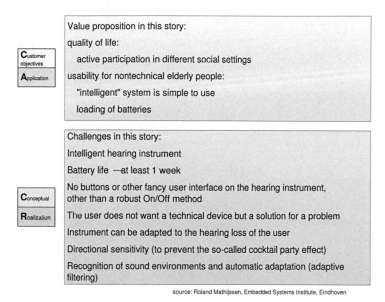

Value proposition in this story:

quality of life:

   active participation in different social settings

usability for nontechnical elderly people:

   "intelligent" system is simple to use

   loading of batteries

Challenges in this story:

Intelligent hearing instrument

Battery life —at least 1 week

No buttons or other fancy user interface on the hearing instrument, other than a robust On/Off method

The user does not want a technical device but a solution for a problem

Instrument can be adapted to the hearing loss of the user

Directional sensitivity (to prevent the so-called cocktail party effect)

Recognition of sound environments and automatic adaptation (adaptive filtering)

source: Roland Mathijssen, Embedded Systems Institute, Eindhoven

Figure 4.31   Value and challenges in this story.

Figure 4.31 shows, for the value and challenge criteria, what this story contributes.

EXERCISES

IN CLASSROOM FOR STUDENTS WITH WORKING EXPERIENCE

Create a story for the product you are working on. The story must fulfill the criteria in Figure 4.29. Make a presentation of the story on a flipchart.

IN CLASSROOM FOR STUDENTS WITHOUT WORKING EXPERIENCE

Create a story for the product provided by the teacher. The story must fulfill the criteria in Figure 4.29. Make a presentation of the story on a flipchart.

PLENARY PRESENTATION AND DISCUSSION

Ask the audience to assess the criteria of Figure 4.29.

5 Strategy

5.1 INTERMEZZO: BUSINESS STRATEGY–METHODS, AND MODELS

5.1.1 INTRODUCTION

The business strategy is input to many activities of architects. Lack of a clear strategy complicates the work of architects. On the other hand, architects need to contribute to the creation and evolution of the business strategy.

The "strategy world" is full of concepts. We will provide a few simple models to position and explain these concepts. There is also an extensive amount of methods and techniques to create and evolve a strategy. We discuss a few methods and techniques that fit in with the architecting contribution. Knowledge of concepts and strategy methods and techniques facilitates architects to contribute to the strategy process.

5.1.2 BASIC CONCEPTS

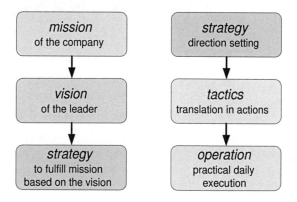

Figure 5.1 Some basic concepts.

Nowadays companies foster an identity of the company by formulating a mission. The mission can be supported by the articulation of four typical company values. The company identity is used for branding: what is the image of the company, and how is the company perceived by the market, its customers, and its shareholders? The mission and company values tend to be very generic, providing a direction to managers and employees. The mission is shown at the top in Figure 5.1.

The leaders in the company formulate a vision: what value can the company bring to the world, what role can the company play? The vision tends to be more market domain specific and will evolve over time.

A true vision is a powerful instrument, uniting the company employees by a shared vision. Unfortunately, too many visions are the result of a mechanistic process. The creation of a vision depends on leaders with the ability to combine a huge amount of context data in a sensible picture. A poor vision might result in ghost hunting or lack of cohesion in the organization.

The mission and vision set the scope for the strategy: where does the company want to go and why? On the right-hand side of Figure 5.1, an often-used layering is shown of strategy, tactics, and operation.

Tactics are an elaboration of the strategy,; how can the strategy be best achieved? For example, do we start with top-of-the-line systems, followed by cost reduced systems, or vice versa?

The operations focus on the execution: get things done. Typically, the operations have a fast heartbeat, where resources and activities are managed continuously and deviations or problems are resolved as soon as possible.

Systems architects will often get the mission and company values as given. They will work using mission and values as guiding principles. Architects might be involved in the creation and evolution of the vision. Systems architects should be involved in strategy creation and evolution. They are typically involved in tactics. A significant amount of the architect's time is spent in the operational aspects of product creation.

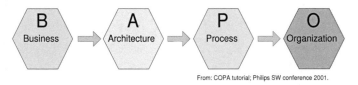

From: COPA tutorial; Philips SW conference 2001.

Figure 5.2 BAPO framework.

Figure 5.2 shows the "BAPO" framework developed at Philips Research by Henk Obbink. Business needs drive the Architecture. Ideally, the Process and Organization should be designed to facilitate the creation of the architecture. In practice, often the opposite is happening: the organizational structure is superimposed on the architecture. In other words, we compromise the architecture to fit in the existing organization. The room for organizational changes triggered by the architecture is limited since organizational changes tend to be slow. The consequence for architecting is that Process and Organization are part of the playing field. We recommend they not be seen as fixed entities, but be taken into account as part of the solution space.

5.1.3 METHODS FOR STRATEGY SUPPORT

SWOT Strengths, Weaknesses, Opportunities, and Threats

One of the methods that is frequently used when creating or evolving a strategy is a SWOT-analysis (see Figure 5.3 for what the letters stand for):

build upon **S**trengths	cope with **W**eaknesses
select **O**pportunities	mitigate **T**hreats

Figure 5.3   SWOT analysis.

Strengths *of the organization, including technology and market position, where the organization can build on.*
Weaknesses *of the organization, where the organization has to cope with these weaknesses. Note that acknowledgment of a weakness and relying on outside support is a legitimate way to cope with weaknesses.*
Opportunities *in the world where the organization can benefit from its current strengths. Opportunities have to be identified, assessed and, finally, a subset has to be selected to pursue.*
Threats *in the world, for example, from changing markets or regulations, or from upcoming competition. Threats have to be identified and assessed, and, when serious, countermeasures need to be formulated.*

The SWOT analysis results in a "big picture" of the current situation, which can be used as starting point for the formulation of a strategy.

Technology Classification

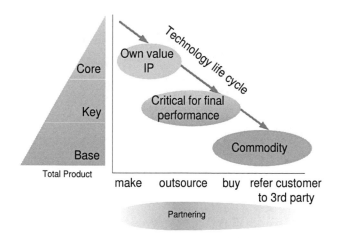

Figure 5.4   Core, key, or base technology.

One of the strategic choices is what a company will do itself and when it will rely on suppliers. There is a spectrum of possibilities, from create and make itself, via outsourcing, to buy. Figure 5.4 shows a technology classification model to reason about these choices. The decision how to obtain the needed technology should be based on where the company intends to add value. The technology classification model uses core, key, *and* base *technology:*

Core *technology is technology where the company is adding value. In order to be able to add value, this technology should be developed by the company itself.*

Key *technology is technology that is critical to the final system performance. If the system performance cannot be reached by means of third-party technology, then the company must develop it themselves. Otherwise, outsourcing or buying is attractive in order to focus as much as possible on core-technology-added value. However, when outsourcing or buying, an intimate partnership is recommended to ensure the proper performance level.*

Base *technology is technology that is available in the market and where the development is driven by other systems or applications. Care should be taken that these external developments can be followed. The company's developments here are de-focusing its attention from core technology. The term Commercial Off The Shelf (COTS) is typically used for base technology.*

5.1.4   EXAMPLES OF STRATEGIC CHOICES

Pay for product

Pay for accessories (cell phone, MP3 cases, skins, etc.)

Pay per use (per printed page, per accessed image)

Pay for service (imaging, printing)

Pay for capability (diagnosis, booklet)

Pay as part of subscription (telecom)

Pay for content (music, movies, eBooks)

Pay for consumables (ink, toner)

Advertizing company pays (Google)

Insurance pays (health care)

Figure 5.5   Examples of business models.

Figure 5.5 shows a list of business models. Every business model has specific characteristics in terms of capital use, return on investment, recurring revenues, variability over time, and margin. On the other hand, the business model will have significant impact on product specification, design choices, organization, staff, and processes.

The position in the value chain is also a strategic choice. Figure 5.6 shows an example of a value chain. Companies that stay at the same position in the value

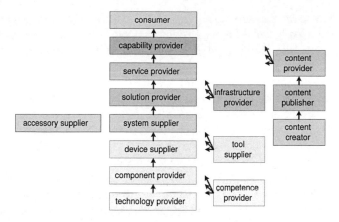

Figure 5.6   Where in the value chain?

chain must protect their margin by excellence in that position. The risk is that "lower" positions in the value chain get a commodity, meaning that the margin gets small or negative. Many organizations address this margin problem by trying to rise in the value chain or by expansion in the value chain.

The choice of the business model and the position in the value chain are primarily business decisions. However, these decisions do have such large impact on the architecting that architects should be involved in the decision making. The consequence for the architects is that they have to participate in a largely financial and economical discussion about the business.

5.1.5   INNOVATION

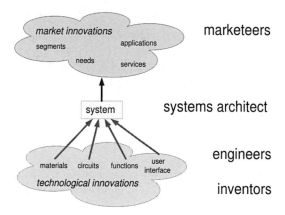

Figure 5.7   Innovation requires all major contributors.

In many organizations, the holy grail of strategy is innovation. Innovation is a *fundamental way to increase the value proposition to the market. Companies have a continuous need for a better value proposition in a world with constant pressure on the margin. The alternative to maintaining the margin at a healthy level is to reduce cost levels.*

Most (mature) organizations achieve the desired improvement of the value proposition by repetitive, small improvement steps. However, many small steps often do not open new markets, or create new applications. Innovation is the result of a creative effort on both the technology side as well as the application and marketing side. Figure 5.7 shows that a concerted effort is needed from truly innovative technology people ("inventors"), engineers, architects, and marketers.

There is a tension between processes and management and innovation. The inherent nature of innovation is to go beyond today's limitations, while processes and management tend to enforce limitations. Innovation requires inspiration rather than control. This same tension can also be observed in the architecting role. Many architects are used to identifying and mitigating risks, a valuable contribution to product creation. However, the risk-based focus can be a severe limitation when searching for innovative solutions.

5.2   ROADMAPPING

5.2.1   INTRODUCTION

The definition of new products is a difficult activity that frequently ends in a stalemate: "It must be done" versus "It is impossible to realize in such a short time frame." The root cause of this frustrating stalemate is most often the fact that we try to solve a problem in a much too limited scope. Roadmapping is a method to prevent these discussions by lifting the discussion to a wider scope: from single product to product portfolio, and from a single generation of products to several generations in many years.

The roadmap is the integrating vision shared by the main stakeholders. A shared vision generates focus for the entire organization and enables a higher degree of cooperating, concurrent activities.

We discuss what a roadmap is, how to create and maintain a roadmap, and the involvement of the stakeholders, and give the criteria for the structure of a roadmap.

5.2.2   WHAT IS IN A ROADMAP?

A roadmap is a visualization of the future (for example, 5 years) integrating all relevant business aspects. Figure 5.8 shows the typical contents of a roadmap. On the right-hand side the owner of the view is shown, while on the left-hand side is shown the asymmetry of the views: the market is driving, while technology, people, and process are enabling.

the key to a good roadmap is the skill of showing the important, relevant issues. The roadmap should provide an immediate insight into the most relevant developments from the five mentioned points of view. These issues are primarily related by

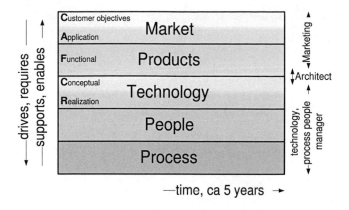

Figure 5.8 The contents of a typical roadmap.

the time dimension.

The convention used in this section is to show products, technologies, people, or process when they are or should be available. In other words, the convention is to be extrovert, to be oriented to the outside world. The introvert aspect, when and how to achieve these items, is not directly shown. This information is often implicitly present since people and process often have to be available before the availability of technology, and technology often precedes the product.

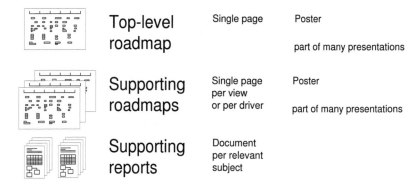

Figure 5.9 The roadmap is documented at several levels of detail.

A good roadmap is documented and presented at top level and at a secondary level with more details. Figure 5.9 shows the desired granularity of the roadmap documentation; the secondary level is called supporting roadmaps. The top level is important to create and maintain the overview, while the more detailed levels explain the supporting data. The choice of decomposition into supporting roadmaps depends on the domain. Typically, the supporting roadmaps should maintain an integrated view. Examples of decomposition are

- One supporting roadmap per key driver
- One supporting roadmap per application area

5.2.3 WHY ROADMAPPING?

The Policy and Planning Process as discussed in Chapter 1, Section 1.1, relies heavily on roadmapping as tool. The main function of roadmapping is to provide a shared insight and overview of the business in time. This insight and overview enables the management of the three other processes:

- Customer-Oriented Process
- Product Creation Process
- People, Process, and Technology Management Process

where managing these processes means defining the charter and the constraints for these processes in terms of budgets and results: Where do we spend our money and what do we get back for it?

When no roadmapping is applied, then the following problems can occur:

Frequent changes in product policy due to lack of anticipation
Late start-up of long lead-time activities, such as people recruitment and process change
Diverging activities of teams due to a lack of shared vision
Missed market opportunities, due to too late a start

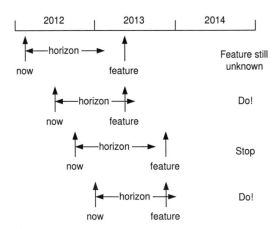

Figure 5.10   Management based on a limited horizon can result in a binary control of product policy decisions.

Frequent changes in the product policy are caused by the lack of time perspective. In extreme cases, the planning is done with a limited time horizon of, for instance, 1 year. External events that are uncertain in time can shift into view within the limited horizon when popular and disappear again when some other hype is passing by. This effect is shown in Figure 5.10

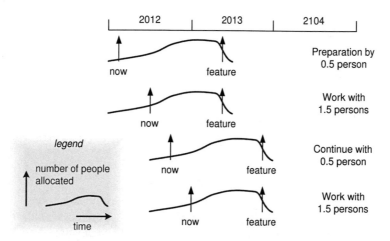

Figure 5.11 Management with a broader time and business perspective results in more moderate control: work with some more or some less people on the feature that is under development for a future release.

The availability of a roadmap will help the operational management to apply a low-pass filter on their decisions. The control becomes more analog rather than discrete, where the amount of people can be increased or decreased dependent on the expected delivery date, as shown in Figure 5.11.

An inherent benefit of roadmapping is anticipation, which is especially important for all long lead-time aspects. Examples are technology, people, and process. This is not limited to development activities only; market preparation, manufacturing, and customer support also require anticipation. For example, reliable mass production has a significant lead-time.

5.2.4 HOW TO CREATE AND UPDATE A ROADMAP

A roadmap is a joint effort of all relevant stakeholders. Typical stakeholders for roadmapping in a typical high-tech company are

Business manager, responsible overall for the enterprise
Marketing managers
People, process, and technology managers, often called line or discipline managers
Operational managers, for example, program managers or project leaders
Architects

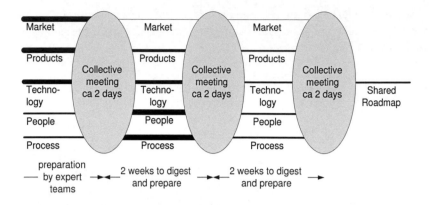

Figure 5.12   Creation or update of a roadmap in "burst-mode."

An efficient way to create or update a roadmap is to work in "burst-mode": concentrate for a few days entirely on this subject. To make these days productive, a good preparation is essential. Figure 5.12 shows the roadmap creation or update as three successive bursts of 2 days.

The input for the first days is prepared by expert teams. The expert teams focus on the *market*, the *products*, and the *technology* layers of the roadmap. The current status of *people* and *process* should be available in presentable format. The target of the first burst is

- To get a shared vision on the market
- To make an inventory of possible products as an answer to the needs and developments in the market
- To share the technology status, trends, and ongoing work as starting point for technology roadmap
- To explore the current status of people and process and to identify the main issues

Between the first and second burst and between the second and third burst, some time should be available, on the one hand, to digest the presented material and the discussions, on the other hand, to prepare the next session. The target of the second burst is

- To obtain a shared vision on the desired technology roadmap
- To share the people and process needs for the products and technology defined in the first iteration
- To analyze a few scenarios for the layers *products*, *technologies*, *people*, and *process*

The thickness of the lines in figure 5.12 indicates the amount of preparation work for that specific part of the roadmap. It clearly shows the shift in attention from the market side in the beginning to the people and process side later. This shift in attention corresponds with the asymmetry in Figure 5.8: the market is driving the business; the people and processes are enabling the business.

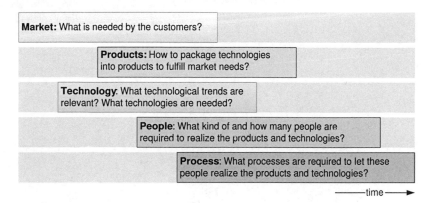

Figure 5.13   Roadmap activities visualized in time.

The function of the collective meetings is to iterate over all these aspects and to make explicit business decisions. The *products* layer of the roadmap should be consistent with the *technology*, *people*, and *process* layers of the roadmap. Note that the marketing roadmap may not be fulfilled by the products roadmap; an explicit business decision can be made to leave market segments to the competition.

Figure 5.13 shows roadmap activities in time. In the vertical direction the same convention is used as in Figure 5.8: the higher layers drive the lower layers in the roadmap. This figure immediately shows that, although "products" are driving the technology, the sequence in making and updating the roadmap is different: the technological opportunities are discussed before detailing the *products* layer of the roadmap.

5.2.5   ROADMAP DEPLOYMENT

The roadmap is a shared vision of the organization. This vision is implemented in smaller steps, for instance, by defining outputs per program and the related resource allocations per program. In Figure 5.14, it is shown that roadmap updates are performed regularly, every year in this figure. After determining the vision, a "budget" is derived that sets the charter for the programs. The budget is revised with an higher update frequency, typically every 3 months. The budget itself sets goals and constraints for the operation. The programs and projects have to realize the outputs defined in the budget in the operation. The operational activity itself uses detailed schedules as means for control. The schedules are updated more frequently than the budget update. Within the operational activity, the updates are mostly event driven: changes in the market, technology, or resources that render the existing plan obsolete.

From long-term vision to short-term realization is a three-tier approach as shown

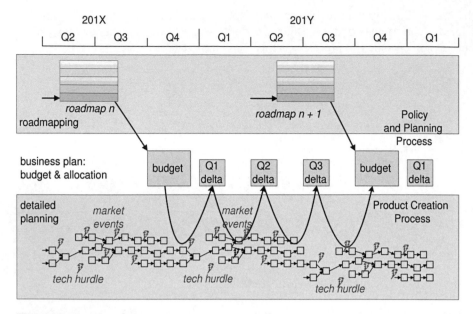

Figure 5.14 The roadmap is used to create a budget and resource allocation. Operational programs and projects use more detailed plans for control.

	horizon	update	scope	type
roadmap	5 years	1 year	portfolio	vision
budget	1 year	3 months	program	commitment
detailed plan	1 mnth-1yr	1 day-1 mnth	program or activity	control means

Figure 5.15 Three planning tiers and their characteristics.

in Figure 5.15. The roadmap provides the context for the budget, and the budget defines the context for the detailed plans. The highest tier, the roadmap, has the longest horizon, the slowest update rate, and the broadest scope. When going down in tiers, the horizon tends to decrease, the update rate increases, and the scope decreases. The roadmap provides a vision and, as such, is not a committment. A budget is a commitment to all involved parties. Plans are means to realize the programs and projects, and tend to adapt frequently to changing circumstances.

5.2.6　ROADMAP ESSENTIALS

We recommend creating a roadmap that fulfills the following requirements:

- Issues are recognizable for all stakeholders.
- All items are clearly positioned in time; uncertainty can be visualized explicitly.

- The main events (enabling or constraining) must be present.
- The amount of information has to be limited to maintain the overview.

Selection of most important or relevant issues

The art of making a roadmap is the selection of the most relevant issues. It is quite easy to generate an extensive roadmap, visualizing all marketing and technological information. However, such superset roadmap is only the first step in making the roadmap. The superset of information will create an overload of information that inhibits the overview we strive for.

Key drivers as a means to structure the roadmap

In Chapter 3, Section 3.3, key drivers are explained as an effective method to elicit and understand needs. They can also be very helpful in the creation and updating of the roadmap. On the marketing side, the trend in these key drivers must be visible in the roadmap. Showing key driver trends also helps to structure the roadmap.

The supporting roadmaps can clarify how the key driver trends will be supported. For instance, a technology roadmap per key driver is a very explicit way to visualize the relationship between the market in terms of key drivers, products with the expected performance levels, and enabling technologies.

Nothing is certain, ambiguity is normal

A roadmap is a means to share insight and understanding in a broader time and business perspective. Both dimensions are full of uncertainties and mostly outside the control of the stakeholders. It cannot be repeated often enough that a roadmap is **only** a vision (or dream?).

The only certainty about a roadmap is that reality will differ from the vision presented in it.

As a consequence, the investment in making the roadmap more accurate and more complete should be limited. Nobody can predict the future; we will have to live with rather ambiguous visions and expectations of the future.

Use facts whenever possible

The disclaimer that *ambiguity is normal* can be used as an excuse to deliver sloppy work. Unfortunately, a sloppy roadmap will backfire on its creators. It is recommended that a roadmap be based on facts whenever possible. Examples of sources of facts are

- Market analysis reports (number of customers, market size, competition, or trends)
- Installed base (change requests, problem reports, historical data)
- Manufacturing (statistical process control (SPC))
- Suppliers (roadmaps, historical data)

- Internal reports (technology studies, simulations)

Use of multiple data sources enables cross-verification of the sanity of the assumptions. For instance, predictions of the market size in units or in money should fit with the amount of potential customers and the amount of money these customers are capable of spending (and willing to spend).

Do not panic in case of impossibilities

It is quite normal that the roadmap layers appear to be totally inconsistent. For instance, a frequently occurring effect is that the budget estimate in response to the market requirements is three times the available budget[1]. Retrospective analysis of past roadmaps shows that the realized amount of work for the given budget is often twice the estimate made for the roadmap. In other words, due to a number of effects, the roadmap estimates tend to have a pessimistic bias. Overestimation can be caused by

- Quantization effects of small activities (the amount of time is rounded to person weeks/months/years).
- Uncertainty is translated into margins at every level (module, subsystem, system).
- Counting activities twice (e.g., in technology development and in product development).
- Quantization effects of persons/roles (full-time project leader, architect, product manager, etc., per product).
- Lack of pragmatism, a more extensive technical realization than required for market needs.
- Too many bells and whistles without business or customer value.

Initial technical proposals might be more extensive than required for market needs, as mentioned in the lack of pragmatism. Technical ambition is good during the roadmap process as long as it does not preempt healthy decisions. The roadmapping discussions should help to balance the amount of technology anticipation with needs and practical constraints.

5.3 INTERMEZZO: CHANGE MANAGEMENT–INTRODUCING SYSTEMS ARCHITECTING ASPECTS

5.3.1 INTRODUCTION

Many organizations do not have explicit roles for systems work or do not use explicit processes, methods, or techniques for systems architecting. There are also many organizations that are unaware of any systems aspects. The introduction of any systems

---

[1] This factor three is an empirical number that of course, depends on the company and its culture.

aspect in an organization is far from trivial. Introducing something new induces a negative reaction not only for systems-related aspects. The field of Change Management addresses the question how to introduce changes into an organization.

Some heuristics from Change Management are

- People do not want to **be** changed. They are quite often willing to change.
- Changing the way of working or changing the culture costs many years.
- It is recommended that multiple tracks be worked at the same time: amongst others, managerial, operational, strategic, etc.
- Changes are better accepted when the initiators earn credit by showing usable results.

5.3.2 EARNING CREDIT, WORK ON URGENT ISSUES

An effective way to introduce changes, such as new systems architecting methods or introducing the role of a systems architect, is to earn credit by actively contributing to the organizational results. Earning credit works fastest when urgent problems are resolved. For example, systems architects typically can contribute to troubleshooting during systems integration. The systems integration phase is always hectic with plenty of time and resource pressure, where monodisciplinary engineers point to other disciplines as the source of problems. The integral overview and the systems thinking capabilities of systems architects make them into ideal troubleshooters. Unfortunately, systems architects not always fancy this "foot in the mud" work.

An approach that nearly always fails is the "evangelism" approach, where systems architects try to convince the stakeholders of the value of new methods or roles by promoting the (theoretical) benefits. Most stakeholders are wary of unproven claims, especially if the messenger has not shown any ability yet.

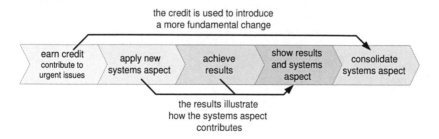

Figure 5.16 Earn credit and work by example.

Spending credit is going faster and easier then earning credit. We recommend keeping on earning credit by working on actual (urgent) issues when introducing systems aspects. Every time that some small change is introduced, architects have used some of their earned credit. Note that forcing changes costs a huge amount of credit, that architects can rarely afford.

Figure 5.16 shows how to introduce changes earning credit, followed by creating an example, and finally by consolidating the change using the credit earned initially. This flow shows that the introduction of the change is done by showing an example rather than preaching change. An example is more easily understood than a theoretical explanation, while the success of the application helps to sell the idea.

5.3.3   EXAMPLE: BOOTSTRAPPING THE ROADMAPPING PROCESS

Many companies and business units have no ongoing roadmapping activity or only a limited one, for instance, a Product Roadmap only. The introduction of a roadmapping process, as described in Section 5.2, is a daunting task for a systems architect. Roadmapping is an improvement at the strategic level with mostly a long-term impact. System architects need to be sufficiently known and respected in an organization to introduce roadmapping; it requires a significant amount of credit to introduce such long term improvement.

Introduction of a roadmapping process can be viewed as part of a change management process. Based on the Change Management heuristics, we recommend introducing roadmapping in a number of smaller steps. The motto here is: Think big, act small.

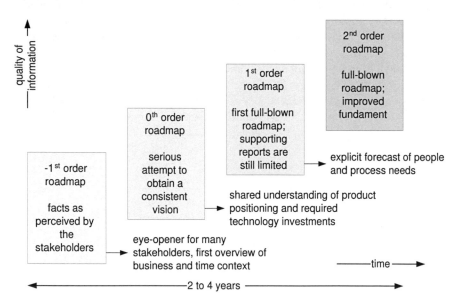

Figure 5.17   Bootstrapping the roadmap process.

Figure 5.17 shows the bootstrapping of a Roadmap Process, typically taking 2 to 4 years. The benefits of starting with roadmapping become available during the first iteration. The mature roadmap, achieved in 2 to 4 years, will bring the full benefits of organizational efficacy and efficiency.

A good start is to capture the existing visions, plans, budgets, etc., and to inte-grate this information into a "minus one"-order roadmap. In most cases, posing such questions forces the stakeholders to reflect on the current status. In many cases, the stakeholders discover that their outlook is rather unbalanced (for instance, the first half year is covered in minute detail, the latter period is fuzzy), or the outlook appears to be totally inconsistent (for instance, marketing has an entirely different expecta-tion than development). Hopefully, the stakeholders get an overview and gain insight into the broader context.

The result of the "minus one"-order roadmap is that the architect gains credit and that the stakeholders are motivated to change somewhat. The stakeholders get ripe to make a next step, for instance, to make a zero-order roadmap.

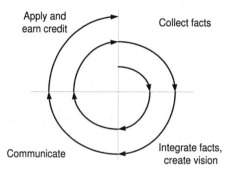

Figure 5.18 Bootstrapping the roadmap process requires a repetition of four steps, as visu-alized by this spiral.

A zero-order roadmap is the first attempt to get the market, the product, and the technology roadmap in place. Such a partial roadmap again helps to earn credit, but it also helps to keep the stakeholders involved. The critical aspect here is the team-building aspect. Roadmapping is a team activity, requiring mutual respect and trust, to enable the open and critical communication needed for the selection of the truly essential issues in the roadmap.

The entire roadmapping process is a repetition of the same activities, visualized in Figure 5.18:

- Collect facts (e.g., market, product, technology)
- Integrate facts and create a vision, where the architect helps in the selection, sim-plification, the interpretation, and presentation.
- Communicate to a broad group of stakeholders in the organization.
- Apply the consequences for the short term and earn credit by showing a positive contribution.

Of course, these four steps are not entirely sequential; they represent the main flow of the process.

5.4 MARKET PRODUCT LIFE-CYCLE CONSEQUENCES FOR ARCHITECTING

5.4.1 INTRODUCTION

A class of products serving a specific market evolves over time. This evolution is reflected in the sales volume of these products. The systems architecting approach depends where products are in this evolution.

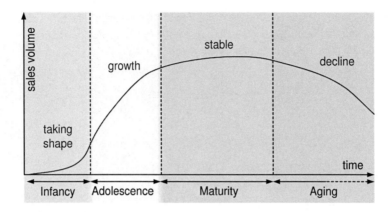

Figure 5.19   Ideal bathtub curve.

The life cycle of a product market combination can be visualized by showing the sales volume as a function of time. In the literature, the form of the curve of the sales volume as function of time is described as bathtub; see Figure 5.19. It is customary to recognize four phases in this curve:

- The life cycle starts with very small sales in the **infancy** phase, where the product finds its shape.
- A fast-increasing sales volume in the **adolescent phase**.
- A more or less stable sales volume in the **mature** phase.
- A decreasing sales volume in the **aging** phase.

The curve and its phases represent theoretical evolution. In the next paragraphs we will discuss observations in practice and an explanation, and we will show that the class of products and the market themselves also evolve on a macro scale.

5.4.2 OBSERVED LIFE-CYCLE CURVE IN PRACTICE

Henk Obbink (Philips Research) observed dips in the sales volume, as shown in Figure 5.20. The transition from one phase to the next does not seem to happen smoothly. In some cases, the sales drops further and the product does not make the transition at all

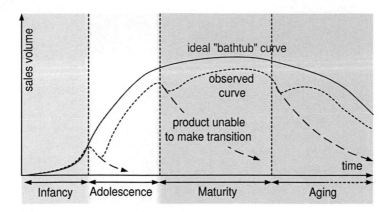

Figure 5.20   Market product life-cycle phases.

The hypothesis for the dips in the curve is that the characteristics of all stake-holders are different for the different life-cycle phases. If the way of working of an organization is not adapted to these changes, then a mismatch with the changed circumstances results in decreasing sales. Figure 5.20 also indicates that, if no adaptation to the change takes place, sales might even drop to zero. No sale will effectively kill the business while still plenty of market opportunity is present.

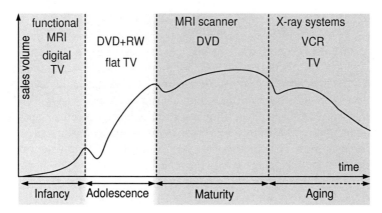

Figure 5.21   Examples of product classes on the curve.

Figure 5.21 annotates the life-cycle graph with a number of products and their positioning in the life cycle. As can be seen, products can move backward in the phases (i.e., become "younger") by the addition of innovative features. For instance, Magnetic Resonance Imaging (MRI) scanners moved backwards when *functional imaging* was added, an innovative way to visualize the activity of specific tissues. Similarly, conventional televisions rejuvenated multiple times by adding digital processing, flat screens, and digital interfaces.

5.4.3 LIFE-CYCLE MODEL

	Infancy	Adolescence	Mature	Ageing
Driving factor	Business vision		Stable business model	Harvesting of assets
Value from	Responsiveness	Features	Refinements / service	Refining existing assets
Requirements	Discovery	Select strategic	Prioritize	Low effort high value only
Dominant technical concerns	Feasibility	Scaling	Legacy Obsolescence	Lack of product knowledge Low effort for obsolete technologies
Type of people	Inventors & pioneers	Few inventors & pioneers "designers"	"Engineers"	"Maintainers"
Process	Chaotic		Bureaucratic	Budget driven
Dominant pattern	Overdimensioning	Conservative expansion	Midlife refactoring	UI gadgets

Figure 5.22 Attributes per phase.

Figure 5.22 shows typical attributes of the life-cycle phases.

The *infancy* phase is characterized by uncertainty about customer needs and, therefore, product requirements. It is essential that the creator/producer be responsive to customer needs, which will provide insight into the needs and requirements. The way of working in this phase reflects the inherent uncertainty, chaotic development, and innovative and pioneering mindset. Product cost is still less of an issue; the risk related to uncertainty is the dominant concern. The design copes with uncertainty by overdimensioning those aspects that are perceived to be the most uncertain.

The *adolescent* phase is characterized by strong (exponential) growth of the sales volume, concurrent with an increase in performance, features, and product variants. The challenge is to cope with this strong growth in many dimensions. With respect to the requirements a strategic selection is needed to serve the growing customer base without drowning in an exploding complexity. The technical and process challenge is to scale up in all dimensions at the same time. Upscaling the Customer-Oriented Processes and the Product Creation Process requires more shared structure between the participants. This involves a mindset change: fewer inventors, more designers. The design pattern used frequently in this phase is the conservative extension of a base design.

The *mature* phase is characterized by more stability of the business model and the market, while the market has become much more cost sensitive. Instead of running along in the feature race, more attention is required to optimize specification and development choices. The value can be shifting from the core product itself to services and complements of the product, while the features of the product are mostly refined. The age of the product starts to interfere with the business, obsolescence problems occur, as well as legacy problems. Innovative contributions become counterproductive; more rigid engineers are preferred above creative designers. Cost optimization is obtained by process optimization, where the processes also become much more rigid but also more predictable, controllable, and executable by a large community of less-educated engineers. The design copes with the aging technology by performing limited refactoring activities in areas where return on investment is still likely.

The *aging* phase is often the phase where the product is entirely seen as a cash cow, maximizing the return on (low) investments. This is done by searching all the low-effort high-value requirements, resulting mostly in small refinements to the existing product. Often integral product knowledge and even specialist knowledge has been lost. Only very important obsolescence problems are tackled. Again, the mindset of the people working on the product is changing to become more maintenance oriented. Cost is a very dominating concern; budgets are used to control the cost. Many changes are cosmetic or superficial, taking place in the most visible parts of the product: the user interface and outer packaging.

EXERCISES

IN CLASSROOM FOR STUDENTS WITH WORKING EXPERIENCE

Create a roadmap for the business where you are working.

IN CLASSROOM FOR STUDENTS WITHOUT WORKING EXPERIENCE

Create a roadmap for the business provided by the teacher.

HINTS TO DO THE EXERCISE

The working order is

1. Market: trends and needs in the outside world
2. Technology: trends and developments outside and inside the company
3. Products: how technology can be packaged to deliver solutions to the market

Use yellow note stickers to populate a roadmap on a flipchart. Make sure that team members write stickers during the discussion.

6 Harvesting Synergy, Product Families

6.1 PRODUCT FAMILIES AND GENERIC ASPECTS

6.1.1 INTRODUCTION

Harvesting synergy between products or projects is being done under many different names, such as shown in Figure 6.1. We use *generic developments* or *harvesting synergy* as a label for this phenomenon. The reader may substitute the name that is used in their organization.

Platform
Common components
Standard design
Framework
Family architecture
Generic aspects, functions, or features
Reuse
Products (in project environment)

Figure 6.1 Different names for development strategies that strive to harvest synergy.

Many trends (increased variability, increased number of features, increased interoperability and connectivity, decreased time to market, globalization of development, and globalization of markets) in the world force organizations into these strategies where synergy is harvested. Harvesting synergy is, however, also a complicating factor both organizationally and technically. We strive to give insight into both needs and complications of harvesting synergy, in the hope that awareness of the complications will help to establish an effective synergy-harvesting strategy.

6.1.2 WHY GENERIC DEVELOPMENTS?

Many people advocate generic developments, claiming a wide range of advantages, such as listed in Figure 6.2.

Effective implementation of generic development has proven to be quite difficult. Many attempts to achieve these claims by generic developments have resulted in the opposite of these claims and goals, such as increased time-to-market, quality and reliability problems, etc. We need a better rationale for generic developments in order to design an effective Shared Assets Creation Process.

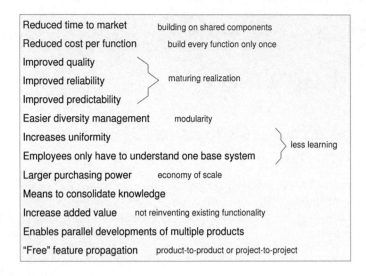

Figure 6.2 Advantages that are often claimed for generic developments.

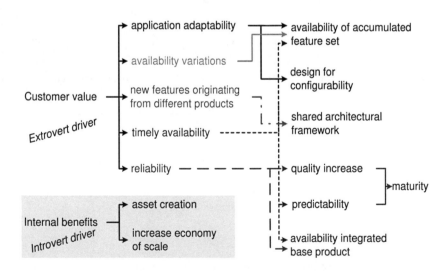

Figure 6.3 Drivers of generic developments.

Figure 6.3 shows drivers for generic developments and the derived requirements for the Shared Assets Creation Process. The first driver (*Customer value*) is extrovert: does the product have value for the customer, and is the customer willing to buy the product? The second driver (*Internal Benefits*) is introvert; it is the normal economic constraint for a company.

Today's high-tech companies are knowledge and skill constrained in a market that is extremely fast changing and rather turbulent. Cost considerations are an economic constraint that has to be balanced with the capability to create valuable and saleable products.

The derivation of the requirements for product development shows that these requirements are not a goal in itself but a means to facilitate a higher-level goal. For instance, a shared architecture framework is required to enable features developed for one product to be used for other products, too. This propagation of features makes sense if it creates value for a customer. So, the verification of the shared architecture framework requirement has to involve the propagation of a new feature from one product to the next, using limited effort and lead-time.

We emphasize the derivation from drivers to requirements because many generic developments fulfill requirements, such as *availability of accumulated feature set*, *designed for configurability*, *shared architectural framework*, and *maturity of implementation*, without bringing in the assumed customer or sales value. For example, many generic developments result in large monolithic solutions without flexibility, but with long development times. Developers of such framework have been providing replies such as: "You cannot have this easy shortcut because our architectural framework does not support it; changing the framework will cost us hundred person-years in three years' elapsed time."

6.1.3 GRANULARITY OF GENERIC DEVELOPMENTS

Granularity is one of the key design choices for systems architects: what is an appropriate decomposition level for modularity? Granularity decisions have to be made at all levels for different purposes, for example, in the application, granularity of functions and roles; at the specification level, granularity of options and features; in conceptual design, granularity of functions and concepts; and in implementation, granularity of many operations.

Figure 6.4 shows the granularity of generic developments in two dimensions. The vertical dimension is the preparation level: What is the intended scope of the generic developments, and how far is the deployment prepared? The horizontal dimension is the integration level: How far are the generic developments integrated when *product developers* deploy generic development?

Both axes range from (atomic) component to (configurable) system. Developments on the diagonal axis, which have a scope where the preparation level is equal to the integration level, are straightforward developments in which integration takes place as far as autonomously possible. Some generic developments concentrate on the generation of building blocks, postponing ("delegating") the integration to the product developer. For rather critical generic developments, the integration of the

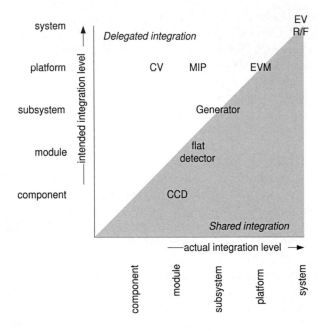

Figure 6.4 Granularity of generic developments shown in two dimensions.

shared asset goes beyond its own deliverable to ensure the correct performance of the asset in its future contexts.

In Figure 6.4 a number of medical generic developments are shown, as an example of categorization.

An extreme example of "delegated" integration is Common Viewing (CV). The medical product devision of a company made an attempt to harvest synergy at the end of the eighties. The vision was to create a large "toolbox" with building blocks that could be used in a wide variety of medical products ranging from Magnetic Resonance Imaging (MRI) scanners to X-ray systems. A powerful set of (mostly SW) components was created, using object-oriented technology and supporting a high degree of configurability.

The CV toolbox proved difficult to sell to product developers, among others, due to the low integration level. The perception of the product developers was that they still had to do the majority of difficult work: the integration. The vision of a marketing manager changed the direction of the CV department into creating a completely integrated product: EasyVision Radiography Fluoroscopy (EV RF). This medical workstation for the URF (Universal Radiography Fluoroscopy) market was highly successful, serving as an intelligent print server. The communication and print function were highly configurable to make the product adaptable to its environment.

The EasyVision RF was used as a basis for a whole series of medical workstations and servers. The shared functionality is developed as generic development at platform level. This platform is nowadays called EasyVision Modules (EVM). Despite its name, it still has a significant integration level, with its upside (product developers are not bothered with the lower-level integration) and its downside (predefined functionality and behavior).

The old CV vision is revived and a second generation of EVM is being created, covering the EVM platform functionality with finer granularity: a module level of integration. The whole evolution as described here from CV as toolbox to more fine grained EVM modules took about 15 years. During all these years, the balance between generality (degree of sharing) and customer value has been changing without ever achieving the combination of a high degree of sharing and a high customer value at the same time.

6.1.4 MODIFIED PROCESS DECOMPOSITION

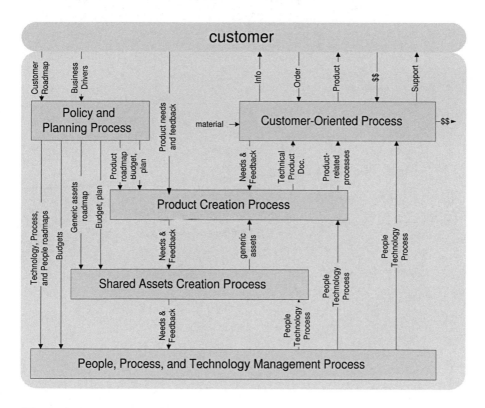

Figure 6.5 Modified process decomposition.

In Chapter 1, Section 1.3, we discussed a simplified process description of companies. This decomposition assumes that product creation processes for multiple products are more or less independent. When generic developments are factored out for strategic reasons then an additional process is added: the Shared Assets Creation

Process. Figure 6.5 shows the (still simplified) modified process decomposition.

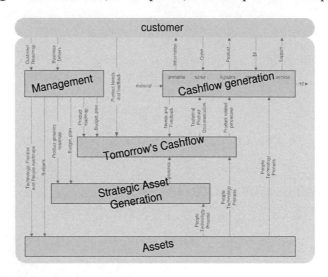

Figure 6.6   Financial viewpoint of processes.

Figure 6.6 shows these processes from the financial point of view. From this point of view the purpose of this additional process is the generation of strategic assets. These assets are used by the Product Creation Process to ensure cash flow for the near future by staying competitive.

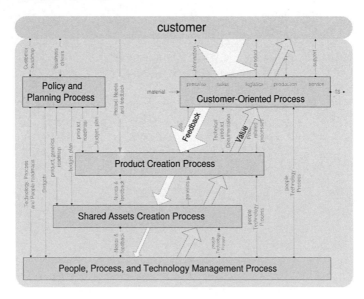

Figure 6.7   Feedback and value flow.

The consequence of this additional process is a lengthening of the value chain and, consequently, a longer feedback chain as well. This is shown in figure 6.7. The increased length of the feedback chain is a significant threat for generic developments. The distance between designers and developers of shared assets and the stakeholders in the outside world is large. These developers easily lose focus on customer value and may focus on the technology instead. Successful sharing requires a strong relation between customer value and technology.

6.1.5 MODIFIED ORGANIZATION OF PRODUCT CREATION

The operational organization of the Product Creation Process is described in Chapter 1, Section 1.3. This organization is a straightforward hierarchy where the limited amount of relations (conflicts) between products or subsystems are managed at the closest hierarchical management level.

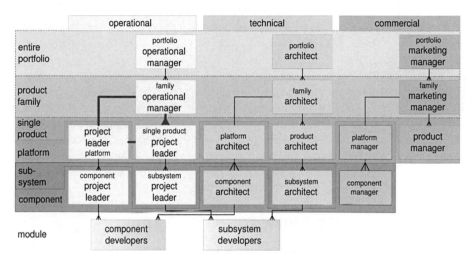

Figure 6.8 Operational organization of the Product Creation Process, modified to enable generic developments.

Introduction of generic developments complicates the operational structure significantly[1]. Figure 6.8 shows the operational organization of the Product Creation Process, with the necessary additions to support generic developments.

The conventional Product Creation Process is based on a relatively straightforward hierarchy, where the control flow and delivery flow are opposite, and where both flows follow the hierarchy. The introduction of generic developments breaks this simple structure: a generic development team delivers to multiple product developments, where the control is taking place from an encompassing operational level to

---

[1]The complication can be avoided by working sequentially. However, in today's dynamic market, sequential work results in unacceptable lead-times. Concurrent engineering is a fact of life. Organizations are looking for opportunities to reduce the lead-time further.

enable operational balancing of products and generic developments. In other words, the principal of the project leader is not the customer anymore but an intermediate manager.

Every operational entity needs the three complementing processes in the product creation process: operational management, design control, and commercial management. For each of these processes, a role is required of someone responsible for that process: the operational manager, the architect, and the commercial manager. Together, these three people form the core team of the operation. Introduction of generic developments also requires the introduction of these roles for the shared assets, such as platform or components.

For the architect role, this means that a platform architect is needed who is closely working together with the platform project leader and the platform manager. On the other hand, the platform architect needs many architectural contacts with the product family architect, acting as the architectural principal, with the product architects acting as customers, and with the component architects acting as suppliers.

The separation of the roles of the platform architect and the product family architect is not obvious. For example, in [9] three operational entities with related processes and roles are identified. Application Family Engineering (AFE), Component System Engineering (CSE), and Application System Engineering (ASE) map respectively on the product family, component, and product as shown in Figure 6.8. We will either have a gap or a double role when mapping four operational entities on three processes. In practice, the result is that one of the roles is missing or played implicit. For instance, quite often, the application family engineer starts to play platform architect, forgetting the original task: *application family* engineering. We have observed that architects either tend to play the platform architect role or the product family role. Architects combining both roles are naturally scarce.

6.1.6   APPROACHES TO GENERIC DEVELOPMENTS

Many different approaches for the development of shared assets are in use. An important differentiating characteristic is the driving force, often directly related to the de facto organizational structure. The main flavors of the driving forces are shown in Figure 6.9.

Lead Customer

The lead customer as driving force guarantees a direct feedback path from an actual customer. Due to the importance of feedback, this is a very significant advantage. The main disadvantages of this approach are that the outcome of such a development often needs a lot of work to make it reusable as a generic product. The focus is on the key functions and performance parameters of the lead customer, while all other functions and performance parameters are secondary in the beginning. Also, the requirements of this lead customer can be rather customer specific, with a low value for other customers.

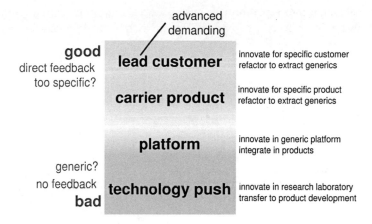

Figure 6.9 Approaches for SW reuse.

Carrier Product

The generic development can be combined with the development of a so-called carrier product to shorten the feedback cycle. The carrier product is one of the ongoing product developments that benefits and suffers from being a representative user of the generic assets. The development of the carrier product benefits from the direct relationship between platform and product, and hence, from the responsiveness from the platform to the product needs. At the same time, the carrier product development suffers from being the early adopter and hitting less mature realizations.

The feedback cycle in a carrier product approach is not as direct as with the lead customer. Combination with a normal product development will result in a better coverage of performance parameters and functionality. the disadvantage can be that the operational team takes full ownership of the product (which is good!) while giving the generic development second priority, which, from family point of view, is unwanted.

In larger product families, the different charters of the product teams create a political tension. Especially in immature organizations or power-oriented cultures, this can lead to counterproductive political games.

Lead-customer-driven product development, where the product is at the same time the carrier for the platform, combines the benefits of the lead customer and the carrier product approach. In our experience, this is the most effective approach in generic developments. A prerequisite for success is an open and result-driven culture to pre-empt any political games.

Platform

In maturing product families, generic developments are often decoupled from product developments by creating an autonomous Shared Asset Creation Process. In products where integration plays a major role (nearly all products), the shared assets are preintegrated into a platform or base product. Platforms or base products follow their own release process before they can be used by product developments.

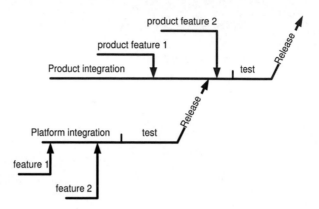

Figure 6.10 The introduction of a new feature as part of a platform causes an additional latency in the introduction to the market.

The benefit of this approach is separation of concerns and decoupling of products and platforms into smaller manageable units. These benefits are also the main weakness of such an approach: as a consequence, the feedback loop is stretched to a dangerous duration. At the same time, the duration from feature or technology to market increases; see Figure 6.10.

Alternative Generic Development Scenarios

A number of alternative reuse strategies have been applied with more or less success:

Spin-off as an independent company is especially tried for key and base technologies. However, many spin-off companies have been reabsorbed by their parent companies. Examples are multimedia processors from TriMedia (parent Philips Semiconductors, later NXP) and cell phone operating system Symbian (parent Nokia).

Reuse after use works quite well in practice, especially for good, clean designs.

Opportunistic copy taking available implementations. The results are quite mixed. Short-term benefits are quick results and hence short feedback cycles. In the longer term, a problem can be that an architectural mess that has been growing can turn into a legacy.

Open source where key and base technologies are shared and developed much more publicly.

Inner-source in the form of a company that stimulates sharing. Sharing within a company is modeled after an open source approach.

Evolutionary refactoring where the architecture and its components are actively refactored to keep them fit for the future and for potentially increased scope of application.

6.1.7 COMMON PITFALLS

We learn from our mistakes. Unfortunately, many mistakes have been made in the area of generic developments. We have compiled a list of pitfalls, as shown in Figure 6.11, from mistakes in generic developments in the past. Some of the attempts to harvest synergy were partially successful, but issues from this list limited the degree of success.

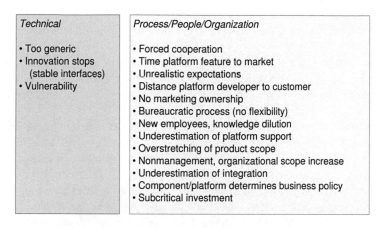

Technical	Process/People/Organization
• Too generic • Innovation stops (stable interfaces) • Vulnerability	• Forced cooperation • Time platform feature to market • Unrealistic expectations • Distance platform developer to customer • No marketing ownership • Bureaucratic process (no flexibility) • New employees, knowledge dilution • Underestimation of platform support • Overstretching of product scope • Nonmanagement, organizational scope increase • Underestimation of integration • Component/platform determines business policy • Subcritical investment

Figure 6.11   Sources of failure in generic developments.

Most of the problems have a root cause in people, process, or organizational issues. The list with technical problems is relatively small:

Too generic platform or components that can do everything but nothing really good: the "Swiss army knife."

Innovation stops because existing interfaces are declared to be stable. Existing structure and interfaces can block innovation.

Vulnerability of all products for common weaknesses is caused by using one and the same core. If the shared core has a problem anywhere, then all products are hit simultaneously. Diversity is a characteristic that enhances resilience. In nature, species often survive disasters, such as diseases, due to the diversity in the population.

The list of people, process, and organizational issues is much longer:

Forced cooperation by upper management may cause demotivation of employees, and social and political tensions in the organization..

Time platform feature to market is caused by stacked-release procedures.

Unrealistic expectations by upper managementare are often caused by claims from architects and engineers of the benefits of harvesting synergy. When less is delivered than promised, then a negative spiral sets in of cost reduction, and hence, the outcome decreases further.

Distance platform developer to customer is explained in Figure 6.7.

No marketing ownership but only engineering push. Marketing support is crucial, since marketing is one of the key players when making decisions about investments. Lack of marketing ownership results in a continuous fight for funding, with starvation of the generic development in the end.

Bureaucratic process and loss of flexibility are often caused by the increased scope of the operation (common components or platform plus derived products) and often require a more formal organization than the individual products used to have. Formalization easily turns into bureaucratic behavior, slowing down the entire organization.

Knowledge dilution can be caused by the hiring of new employees. Often, an increase in resources is needed early during the creation of shared assets. If these new resources are inexperienced, then the knowledge is diluted, resulting in reduced quality of the created assets.

Underestimation of shared asset support required when the shared assets are used by products. Product designers need support when specifying and designing new products based on these assets, and they need support for troubleshooting during integration and introduction in the field. When components are used in new circumstances (e.g., new products), then always unexpected problems pop up.

Overstretching of product scope is caused by going beyond the natural level of synergy. Harvesting synergy is a balancing act between maximum value creation for specific customers and minimizing diversity in realization. When minimization of diversity dominates over value creation, then customers are not served well, resulting in loss of business. Organizations easily lose their customer focus when creating a synergy drive.

Nonmanagement of organizational scope increase that is inherent when multiple products share assets. The scope increase requires organization, process, and staffing adaptations.

Underestimation of integration of shared assets in other products. Systems integration is often ill understood and hence underestimated; see Chapter 7, Section 8.2. When existing products have to migrate to the use of shared assets, then this requires that these products adapt their architecture, too.

Component/platform determines business policy , which, effectively, is an inversion of the need-driven approach. This inversion relates to the distance between shared asset development and customers. What happens is that what *can* be done dominates over what *needs* to be done. The shared asset developers get de facto power since all products depend on their delivery.

Subcritical investment, caused by a cost reduction focus. Shared asset development primarily should bring market and customer value while keeping the cost limited by harvesting synergy. As soon as cost reduction dominates over value creation, then all products and shared assets can get too little investment, causing delays and quality problems.

6.2 A METHOD TO EXPLORE SYNERGY BETWEEN PRODUCTS

6.2.1 INTRODUCTION

We can distinguish two types of situations where we can strive to harvest synergy, as illustrated in Figure 6.12.

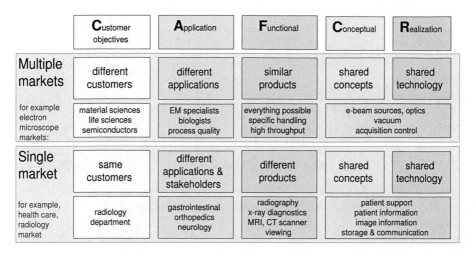

	Customer objectives	**A**pplication	**F**unctional	**C**onceptual	**R**ealization
Multiple markets	different customers	different applications	similar products	shared concepts	shared technology
for example electron microscope markets:	material sciences life sciences semiconductors	EM specialists biologists process quality	everything possible specific handling high throughput	e-beam sources, optics vacuum acquisition control	
Single market	same customers	different applications & stakeholders	different products	shared concepts	shared technology
for example, health care, radiology market	radiology department	gastrointestinal orthopedics neurology	radiography x-ray diagnostics MRI, CT scanner viewing	patient support patient information image information storage & communication	

Figure 6.12 Types of synergy.

The company serves **single markets with different products**. The customer world is homogeneous and needs products that can be quite heterogeneous in both the used concepts and technologies.

The company serves **multiple markets with quite similar products**. The customer world is heterogeneous and needs products that are different but quite similar. The similarity in the products suggests that synergy is present and can be harvested.

Figure 6.12 shows one example in both categories. The radiology department in health care is an example of a homogeneous market, where many different products are interoperating to provide the desired capabilities. Some of these systems are diagnostic equipment with different imaging modalities, for example, x-ray systems, Magnetic Resonance Imaging, and Computer Tomography. However, administration, viewing, communication, and archiving systems are information technology oriented. These different imaging systems have some functionality that is quite similar, and hence, there might be some result out of synergy opportunities.

The market of electron microscopes is an example of a heterogeneous market. Different laboratories in material sciences, life sciences, and semiconductors use very similar imaging technology to study nanometer-scale details. The interests and skills of the operators in the different domains vary widely.

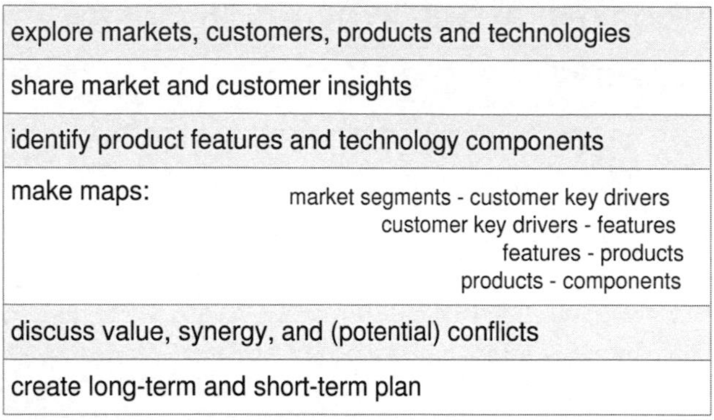

explore markets, customers, products and technologies
share market and customer insights
identify product features and technology components
make maps: market segments - customer key drivers customer key drivers - features features - products products - components
discuss value, synergy, and (potential) conflicts
create long-term and short-term plan

Figure 6.13   Approach to platform business analysis.

6.2.2   STEPWISE METHOD TO EXPLORE SYNERGY OPPORTUNITIES

Figure 6.13 shows the stepwise method to explore and analyze opportunities to harvest synergy. Every step is elaborated more later in this section.

Explore markets, customers, products, and technologies to create a shared understanding of the playing field.

Share market and customer insights by studying one customer and one product, followed by a more extensive study of workflows. Select a demanding customer and a product under discussion as starting point. Note that other products and customers will be studied in later iterations.

Identify product features and technology components by doing initial specification and design work.

Make maps where the views that resulted from the first steps are related and visualized.

Discuss value, synergy, and (potential) conflicts to get the main issues on the table in a factual way.

Create long-term and short-term plan to transform what can be done into something that (probably) will be done.

The whole process described by this method should be performed by an exploration team, a small team of key people, including marketing managers, architects, and key technology experts.

Explore markets, customers, products, and technologies

The exploration is performed by using time boxes with predetermined duration to discuss the following questions by the exploration team:

- What markets do we want to serve?
- What specific customers do we expect? What are the key concerns per customer?

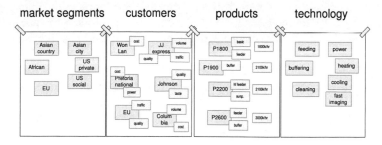

brain storm and discuss time-boxed

Figure 6.14 Explore markets, customers, products, and technologies.

- What products do we foresee? What are the key characteristics of these products?
- What technologies do we need?

The purpose is to make a quick scan of the playing field so that a shared insight is created between the members of the team. Figure 6.14 shows the typical result of the exploration: a number of flipcharts with sticky notes. This first scan can be done in a half day to a full day.

Share market and customer insights

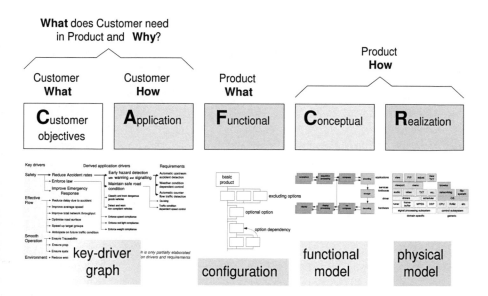

Figure 6.15 Study one customer and product; examples of views are shown in this figure.

The challenge is to get more substance after the first quick scan. Figure 6.15 shows how the CAFCR model is followed to explore one product for one market in more depth. The idea is that one such depth probe helps the team to get a deeper understanding that can be extended to other products by variations on a theme. In this figure, the CAF-views are covered by a key driver graph, the F-view focuses on the required commercial product structure, the Conceptual view is used for a functional model of the system internals and the R-view shows a block diagram. This is an example of a CAFCR analysis, but specific markets and products can benefit from other submethods in the CAFCR views.

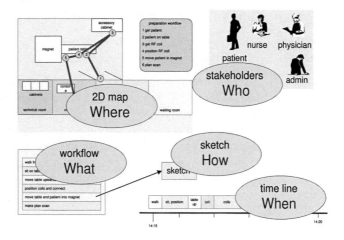

Figure 6.16   Workflow analysis for different customers and applications. The figure shows example of views that can be made for such analysis.

The next step in digging in deeper is to explore the workflow of different customers. Figure 6.16 shows the different perspectives on customer workflow:

Where are workflow steps performed?
What is done in the workflow?
Who is involved?
When are steps performed, and what is the duration?
How are selected steps performed?

A specific insight into the workflow of different customers and applications is critical for later choices about synergy. This step is too often skipped, either because of time pressure or because of ignorance. Insufficient understanding of the use of the systems compromises the products and hence degrades the value for classes of customers. In this step, we try to bridge the marketing gap and the context gap as shown in Figure 2.19 in Chapter 2.

At this moment in the exploration process, the exploration team has insight into different customers. It helps the team and its stakeholders if the growing insight of these different customers and their needs for products can somehow be captured in a

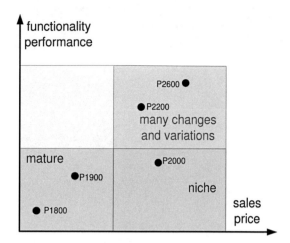

Figure 6.17 Make map of customers and market segments.

single map with a few main characteristics. Figure 6.17 shows a simplistic example. Often, characteristics such as price and performance parameters are used for such a map.

Identify product features and technology components

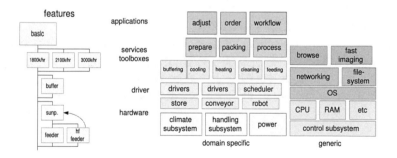

Figure 6.18 Identify product features and technology components.

In this step, the commercial structure of the product is further elaborated: What are the required commercial configurations, what should be optional? Also construction decomposition is elaborated: What are the expected hardware and software components or building blocks, and what are the dependencies between them? The main purpose of this step is to understand the potential commercial and technical modularity. From this modularity, synergy can emerge between products (Figure 6.18).

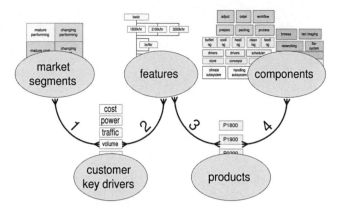

Figure 6.19   Mapping from markets to components.

Make maps

The first views have resulted in the identification of *market segments, customer key drivers, features, products,* and *components*. In this step, the objective is to relate these views, for example, *market segments* to *customer key drivers, customer key drivers* to *features, features* to *products,* and *products* to *components*; see Figure 6.19. Each mapping can be many to many, for example, different market segments can share the same key drivers while every market segment has multiple key drivers.

Discuss value, synergy, and (potential) conflicts

In general, the wish list for features is longer than can be implemented in the first releases. We need more insight into the value of the different features to facilitate a selection process, as discussed in Chapter 3, Section 3.4.

   Figure 6.20 shows the results of this selection process. Note that the discussion provides most of the value to the exploration team. The team is forced to compare features and to articulate their value by the need to characterize and agree on the scoring.

Create long-term and short-term plan

Practical constraints such as time and effort often determine our choices in synergy and the order in which we realize these choices. The exploration team has to translate its vision into a plan showing in what order it could be realized. Part of the plan will be short term: what do we do rather concrete in the next few weeks or months? The long-term plan visualizes the big picture of moving toward synergy: how do we envision that we will migrate to the synergetic situation? Note that, again, making the short-term and long-term plan serves the purpose of forcing the exploration team into this practical discussion.

— products →

| | P1800 | | | P1900 | | | P2200 | | |
	satisfaction customer	sales price	market share	satisfaction customer	sales price	market share	satisfaction customer	sales price	market share
feeder	1	5	4	3	4	4	4	5	5
hf feeder									
buffer	4	3	4	5	3	4	4	3	4
sunpower	2	2	1	2	2	1	2	2	4

features →

Figure 6.20 Determine value of features.

6.2.3 EXAMPLE OF SYNERGY

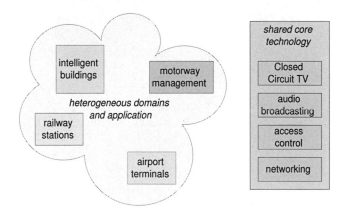

Figure 6.21 Example of synergy between heterogeneous markets.

Figure 6.21 shows an example of a company serving four heterogeneous markets: intelligent buildings, motorway management systems, airport terminals, and train stations. This company performs projects in these four markets, providing closed circuit televisions, access control, audio broadcasting, and the integration. The synergy is in the technical components such as cameras and loudspeakers. The question was if there is also potential synergy in the integration, for example, the networking, system control, and operator interfaces. For that purpose, the key driver diagram (see Chapter 3, Figure 3.10) was developed.

The outcome of this exploration is that it makes sense to share lower-level networking, system control, and operator interfaces. However, the domains may need specific adaptations to be done as a project.

EXERCISES

IN CLASSROOM FOR STUDENTS WITH WORKING EXPERIENCE

Make an inventory in your business where synergy is harvested, where an attempt is made to harvest synergy, and where opportunities for synergy are not harvested.

Assess the success of harvesting synergy and identify success factors and blocking factors for harvesting synergy.

Present the results in one flipchart.

IN CLASSROOM FOR STUDENTS WITHOUT WORKING EXPERIENCE

Answer the following questions for the business provided by the teacher:

1. Identify opportunities of harvesting synergy.
2. Propose a technical implementation to support the synergy.
3. Identify process and organizational issues that need to be solved to make the synergy harvesting successful.
4. Present the results in one flipchart.

7 Supporting Processes

7.1 SYSTEMS ARCHITECTS AND SUPPORTING PROCESSES

7.1.1 INTRODUCTION

In Chapter 1, Section 1.1, we discussed a highly simplified decomposition of a business in processes. Figure 7.1 shows a number of the *supporting processes* as an overlay of Figure 1.1.

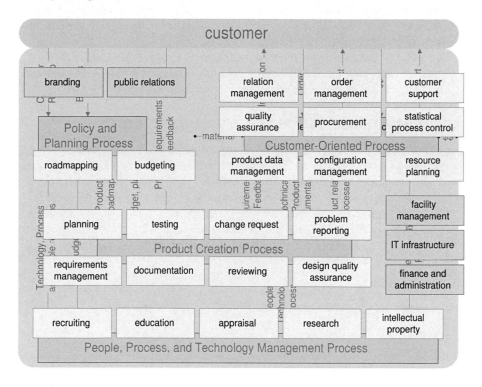

Figure 7.1 Supporting processes overlaid on the simplified decomposition of a business.

These supporting processes are loosely mapped on the main processes. However, many of these supporting processes are more cross boundary than suggested by this diagram. For example, the People, Process, and Technology Management Process is mainly managing intellectual property, but intellectual property also plays a significant role in the Product Creation Process.

In established organizations, these supporting processes tend to be mature; they are well defined and ingrained in the way of working. Normally, these processes evolve over time, following needs and, for instance, tool developments. However, the processes do not always fit the current situation. In practice, the following situations can be observed:

Not sufficient for current situation because product creation challenges have evolved faster than the processes and tools in the company.

Overconstraining or slowing down the product creation work. Processes tend to grow and become heavier over time. The rationale behind control measures is invalid, but nobody is correcting the situation.

Systems architects are often confronted with the consequences of less fit supporting processes; architects see the symptoms of problematic processes, and their work suffers from these problems. If architects do not recognize the root cause of the problem, then they tend to look for solutions in their own domain: systems specification and design. However, the root cause, the failing process, needs to be addressed to solve the more fundamental problem. Solving process shortcomings is not part of the systems architect's role. We will discuss the role of the systems architect in addressing process shortcomings in the next paragraph. In other sections and papers on www.gaudisite.nl, we discuss some of the most common problems in supporting processes, such as documentation, reviewing, and integration.

7.1.2   THE CRITICAL ROLE OF THE SYSTEMS ARCHITECT

Systems architects often detect problems in supporting processes early because they encounter its consequences in dealing with other stakeholders or in executing the prescribed procedures. For example, many organizations prescribe many pages of overhead information in their documentation procedures and templates. Systems architects need to fulfill all overhead, wasting valuable time, and their readers often do not have or take the time to search for the actual contents. In this example, the good intent of the procedures and templates backfires: they do not support product creation, but rather constrain it.

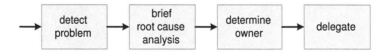

Figure 7.2   From problem detection to allocation.

Systems architects should not resolve supporting process issues themselves. That is often outside their competence and part of the responsibility of others. Figure 7.2 shows the steps that an architect can follow in case of poor supporting processes:

Detection of the problem itself by observing the symptoms.
Perform a brief root cause analysis to ensure that the problems and their causes are well understood.
Determine the owner of the problematic supporting process.
Delegate the solution to the owner of the supporting process. The owner is responsible to improve the process.

The architect is one of the stakeholders (and a customer) of the supporting processes. The process owner ought to take detected problems and stakeholder needs seriously.

Note that the architect should not push a solution. Pushing a solution is overstepping the boundary of the process owner, which often causes a negative reaction.

Systems architects need to find a balance between acceptance of existing procedures and their own need to have appropriate supporting processes. Many architects are too lenient, accepting the burden of poor supporting processes without taking action. On the opposite side are the systems architects who start to reform company processes, outside their own area of competence. The main risk of architects performing process redesign is that the actual architecting work does not get performed. The recommended way is to be critical about the fitness of supporting processes and to communicate shortcomings with the process owners.

7.2 GRANULARITY OF DOCUMENTATION

7.2.1 INTRODUCTION

Documentation is an important means of communication in the Product Creation Process. Multiple authors with different competencies write the whole documentation set. Systems architects contribute to the structure of the documentation and write a small subset of the documentation themselves. The size of the units within the documentation structure is called the granularity of the documentation.

The right level of granularity improves the effectiveness of the documentation. We discuss criteria to design the documentation structure, documentation granularity, and documentation processes.

7.2.2 STAKEHOLDERS

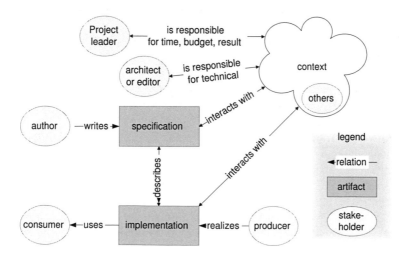

Figure 7.3 The stakeholders of a single document.

Figure 7.3 shows the stakeholders of a document. The document is a description of a function or component that has to be realized by means of an implementation. The producers and consumers of the function or component are the main stakeholders of the document. The author is also an important stakeholder. The function or component is always realized and used within a broader context. This context interacts with the function or component, so the persons responsible for the context are also stakeholders of the document. In this context there will be other stakeholders as well, such as people who are involved with the function or component.

Example–digital flat screen TV

An electronics designer writes a specification for a Printed Circuit Board (PCB) to be used in a digital flat screen TV. A digital designer and a layout engineer realize the design, and hence, they are the producers. A software engineer will write the software, making use of the functionality of the board, and he is one of the consumers. The product (the digital flat screen TV) is the context for this PCB. The designer of the power supply might be a stakeholder, especially if the PCB has specific power requirements. The industrial designer responsible for packaging is another stakeholder. The final product will have a project leader, responsible for schedules, costs, etc., and is a stakeholder with respect to these issues. Finally, the architect is responsible for a balanced and consistent product design, where the PCB should fit in.

7.2.3   REQUIREMENTS

The documentation of a product needs to be decomposed into smaller units, with the smallest units being atomic documents. We will discuss the requirements for the entire documentation structure, the documents themselves, and the underlying process.

The criteria for the entire documentation structure and process are

Accessibility for readers. The information should be understandable and readable for the intended audience. The signal-to-noise ratio in the document must be high; information should not be hidden in a sea of words.

Low threshold for readers. No hurdles such as many pages of meta-information, cumbersome security provisions, or complicated tools should dissuade readers from actually reading the document.

Low threshold for authors. Authors have to be encouraged to write. Hurdles, such as poor tools or cumbersome procedures, provide an excuse to delay the writing.

Completeness of important information. Note that real completeness is an illusion; there are always more details that can be documented. All crucial aspects have to be covered by the entire documentation set.

Consistency of the information throughout the documentation. Writers strive for consistency, but we have to realize that, in a complex world with many stakeholders, some inconsistencies can be present. Inconsistencies that have significant impact on the result have to be removed.

Maintainability of the entire documentation. The organization needs to maintain documentation both during product creation and the rest of the product life cycle.

Scalability of the documentation structure to later project phases. Many more engineers can be involved in the later phases. The following measures help to achieve scalability:

- Well-defined documentation structure
- Explicit overview specifications at higher aggregation levels
- Recursive application of structure and overview documents
- Distribution of the review process

Evolvability of the documentation over time. Most documentation is reused in successive projects.

Process to ensure the quality of the information. The quality of the content of the information is crucial to good results. Documentation that has been made only to satisfy the procedure is a waste of effort and time.

From the reader's point of view, this translates into the requirements for the document infrastructure: it must be fast and easy to *view* and *print* documents, and *searching* in the documentation has to be fast and easy. Searching must be possible in a structured—for example, hierarchical—way and via free text "a la Google." Any part of the documentation must be reachable within a limited number of steps, so no excessively deep document hierarchies are necessary.

The criteria for the documents within the documentation structure are

High cohesion within the document; the information in a document has to "belong" together. If information is not connected to the rest of the document, then this information might belong in another document.

Low coupling with other documents; some coupling will be present since the parts together will form the system. If the coupling is high, then the document decomposition is suspect and might need improvement.

Accessibility for the readers. See accessibility for the entire documentation.

Low threshold for the reader. See low threshold for readers for the entire documentation.

Low threshold for the author. See low threshold for authors for the entire documentation.

Manageable steps to create, review, and change documents. Documents in product creation are reviewed and updated frequently. Hence, these operations should take limited effort and time. The consequence is that single documents should not be large.

Clear responsibilities, especially for the content of the document. Documents with multiple authors are suspect since responsibility for the content can be diffuse. Worse are documents of which an anonymous team or committee is "the author." If a document needs multiple authors, then it is often a symptom of bad decomposition. In addition, the reviewers' responsibility must be clear and hence, we recommend limiting the number of reviewers. When many reviewers are needed, then the decomposition is again suspect.

Clear position and relation with the context since documents only make sense in the intended context. On purpose, the information is captured in multiple documents. Therefore, for every individual document, it should be clear in what context it belongs and how it relates to other documents.

Well-defined status of the information is essential because documents are used and most valuable in the period when they are created. The content can be quite preliminary or a draft. The document must clearly indicate what the status is of its content, so that readers can use it with proper precautions.

Timely availability of the document is important because when documents are available too late, we do not harvest the value. Authors have to balance quality, completeness, and consistency against the required effort and time.

A very important function of documentation is communication. Communication requires that the information is accessible to all stakeholders and that the threshold to produce documentation or to use documentation should be low[1].

7.2.4   DOCUMENTATION STRUCTURE

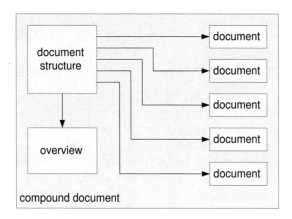

Figure 7.4   Large documents are decomposed into smaller documents, supported by a document structure and overview.

The standard way to cope with large amounts of information is to decompose the information into smaller parts. The decomposition of the large amount of information results in a set of smaller documents. The structure of such decomposition is made explicit in the "documentation structure," fulfilling the requirement to have a *well-defined documentation structure*. The documentation structure is managed as a

---

[1] Quite often, organizations focus on documentation procedures and documentation management, forgetting the main drivers mentioned in this subsection. The result can be tremendous thresholds, causing either apathy or bypasses. It cannot be stressed enough that procedures and tools are the **means** to solve a problem and not a goal in itself.

normal document. An overview document is required to keep the overview accessible, addressing the requirement to have *overview specifications at higher aggregation levels*. Overviews help the readers, especially when the more detailed information gets scattered in smaller documents.

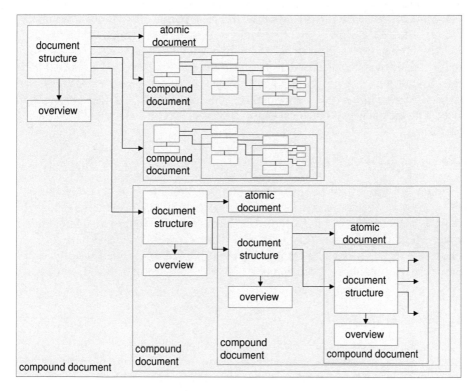

Figure 7.5 Decomposition is applied recursively until the atomic documents fulfill the requirements in Subsection 7.2.3.

This decomposition is applied recursively; see Figure 7.5. In this way, the granularity supports the realization of the requirements as described in Subsection 7.2.3. For instance, the principle of *recursion* is a good answer to the requirements related to *scalability* of the entire documentation. The creation of explicit structure and overview documents and the allocation of creation and maintenance to authors support *maintainability*.

A fine grain structure, for example, small documents, lower the threshold to make documents and read the contents, in this way answering document requirements *accessibility for the reader, low threshold for the reader,* and *low threshold for the author*.

The clarity and value of the content is the foremost requirement for documentation. Decomposing the documentation is a balancing act in many dimensions, similar to the decomposition of systems. The clarity and value of the content may not suffer due to the structure. Dogmatic structuring rules might be conflicting with clear re-sponsibilities (*single author*). When authors write outside their expertise area, then there is a severe quality risk. The decomposition has to result in sufficiently small documents to support the requirement of *manageable steps to create, review, and change*. Large, monolithic documents violate this requirement.

Document granularity is an important design criterion for the documentation structure. The extreme that every *single value* is an entity[2] is not optimal because the relations between values are even more important than the value itself. In the case of *single value* documentation, relations are lost. The other extreme, to put everything in a single document, conflicts with many of the requirements such as *manageabil-ity*, *clear responsibilities*, *well-defined status*, and *timely availability*. The granularity aspect, with the many psychological factors involved, is further discussed in Subsec-tion 7.2.5.

7.2.5 PAYLOAD, THE RATIO BETWEEN OVERHEAD AND CONTENT

An atomic document must be small enough to be accessible to readers. Thick doc-uments are put on top of the stack of "interesting papers to be read," to be removed when this stack overflows. For most people, time is the scarcest resource. Struggling through all kinds of overheads is a waste of their scarce and valuable time. Docu-mentation effectively supports communication if the reader can start directly with reading the relevant information. Figure 7.6 shows the layout of a good document.

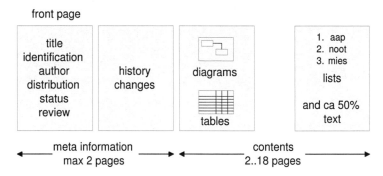

Figure 7.6 Layout of a good document; a heuristic for the number of pages of a good document is $4 \leq number\ of\ pages \leq 20$.

The front page is used for all relevant meta-information. Meta-information is in-formation required for document management, defining the status, responsibilities, context, etc. The history and change information on the second page should be a

---

[2]A common pitfall is to store all values in a database. In this way, every value is an entity in itself. Such a database creates the suggestion of completeness and flexibility, but in reality, it becomes a big heap where designers lose the overview. These databases may help the verification process but do not fulfill documentation needs.

service to the readers to enable them to see quickly the relevant changes relative to earlier versions they might have read. More extensive change information, required for quality assurance purposes, can be present in the document management system; it should not distract the reader from the information itself.

Such a document needs only to be opened to access the contents. Many older organizations tend to make documents with up to 10 pages of overhead information. Many people are interrupted by phone, calendar, e-mail, or person before reaching page three. The overhead de facto inhibits people from reading the contents of badly written documents.[3]

The contents of a well-written document ought to be optimized to get the essential information transferred. The reader community consists of different people, with differing reading and learning styles. To get information across it must be visualized (diagrams), structured, and summarized (tables and lists) and, to a limited extent, explained in the text.

Once a document starts its life cycle, the next risk is that it keeps growing. Authors have the tendency to transform comments and critiques of readers in explaining the text. Unfortunately, large sections of text hide the key information, and violation of the maximum of 20 pages becomes probable. It is better to translate the comments and critiques back into an improved diagram, table, or list. Authors have to find the root cause of reader comments. For example, an unclear diagram gives rise to misunderstanding.

Another frequently occurring trap is the extension of a document with missing context information. For instance, if the higher-level specification is missing, parts of that specification are included in the lower-level specification. An effective counter-measure for this trap is to write the specification structure, showing the context and enabling the writing of the context later gradually. This strategy results in documents that are more focused, have a better cohesion internally, and have less coupling with other documents.

The heuristic mentioned in Figure 7.6 is that a good document should have four or more pages. This minimum should prompt people with the question whether the information in a very small document has a right of existence on its own. The ratio of overhead versus payload for very small documents is unbalanced. There are small documents were the small size is appropriate.

The maximum number of pages for a good document is 20. These documents do not scare people away yet. A 20-page document can be read in less than 1 hour, and the review can be done in the same time. For many purposes, 10- to 15-page documents are optimal. If documents require more than 20 pages, the recipe is simple: make it a compound document, and split the content in multiple smaller documents.

A natural split-up is often directly visible in large documents.

Large documents often violate a number of the requirements in Subsection 7.2.3. For instance, a single person is editing the document, but multiple authors write parts

---

[3]Often the situation is much worse than described here. In name of "standardization," these counter-productive layouts are made mandatory, forcing everyone to create thresholds for readers!

of the document. Another symptom of requirement violation is a document that is partly finished and partly in draft status (for instance, the "requirements" sections are written, while the "design" is still under way).

7.3 INTERMEZZO: LEAN AND A3 APPROACH TO SUPPORTING PROCESSES

7.3.1 INTRODUCTION

LEAN manufacturing is a manufacturing approach based on Toyota's successes, as described by researchers who observed and analyzed Toyota. LEAN product development is building on LEAN manufacturing, where the ideas from the repetitive production environment are transformed for use in the creative product development environment. Likewise, LEAN product development is based on observing Toyota product development.

A popular way to explain LEAN is that the basic principle is to "avoid waste." For the purpose of this chapter, we characterize LEAN by the following elements, loosely based on [12]:

A holistic, systems approach to product development, *including   people,   pro-cesses, and technology*
Multidisciplinary *from early start, with a drive to be fact-based*
Customer understanding *as the starting point*
Continuous improvement and learning *as cultural value*
Small distance *between engineers and real systems, including manufacturing, sales and service, and the system of interest*

7.3.2 LEAN AND SUPPORTING PROCESSES IN GENERAL

LEAN product development delegates responsibilities as much as possible to the experts. The management facilitates and stimulates the experts to operate toward the goals using LEAN principles. The way of working is highly pragmatic, where the goal dominates over the means. In many cases, no complicated computer tools and repositories are used.

Colocation in a larger room is common. The team visualizes plans and schedules on wall space in this room, with low-tech means such as paper, pens, or magnetic boards. The components or a prototype of the system can be present in the room (keep the distance small!).

In LEAN manufacturing and LEAN product development, A3's are used to document and communicate, as discussed in the next section. A3 is a European standard paper size of 297 × 420mm.

7.3.3 A3 ESSENTIALS

An A3 contains a "human-friendly" amount of information. The size permits some depth and facts in the information, while at the same time it forces the author of the

A3 to select and process the information carefully.

We have the following guidelines, loosley derived from [4], when using A3's as the unit of documentation and communication:

Capture "hot" topics *that are currently under discussion; when topics are under discussion, then explicit diagrams facilitate the discussion. The active use of the A3 will stimulate the evolution of the A3 itself.*

Visualize one topic per A3 *so that every A3 is homogeneous. The requirements for documents, defined in Subsection 7.2.3, also apply for A3 documents.*

Show multiple related views. *The strength of the A3 format is that several diagrams can be shown at the same time. These diagrams are different views on the same topic. These views will be related. These relations should be present in a supportive, nondominating, way, for example, by the use of colors, shapes, lines, labels, or naming conventions.*

Make the A3 digestible *by limiting the amount of content. Note that the size limitation forces the authors to limit the amount of information.*

Make the diagrams and information specific; *for example, by showing and quantifying use cases. Note that the risk of the size limitation is that too "empty" or too glossy posters are made. Good A3's have substance; specific information helps to make the A3 substantial.*

Use practical visualizations *close to the experience of the stakeholders. Good A3 documents engage the stakeholders helped by instant recognition of the visualizations.*

Note that the granularity and structuring guidelines of Section 7.2 are applicable to A3-based documentation as well, where the payload size is limited by the A3 dimensions.

7.3.4   EXAMPLE OF AN A3

Figure 7.7 shows an example of an architecture overview on a single A3. This A3 shows the "super-super" system: the wafer back-end factory where nearly finished wafers are processed and where integrated circuits (ICs) are produced. Part of the process at the factory level is the metal printing. Metal printing related process steps are shown at factory level, both visually as workflow steps as well as by quantifying the throughput in minutes per wafer.

The next layer in Figure 7.7 shows the "super" system: the cell. All equipment in factories related to a process step is organized in cells. A cell is a self-sustained unit in the factory that can perform all operations required for this specific process step. The core entity of the cell is the wafer-handling robot. This robot transports wafers from the containers with wafers (so-called FOUPs) to the functions in the cell, such as prefill, clean, and print. The flow of the wafers through the cell is visualized on the right-hand side for one master and one wafer. The previous and next wafers are simultaneously in the cell; the cell is processing wafers in the pipeline mode.

The third layer shows the decomposition of the metal printer, the system of interest. These subsystems are shown as backside and front-side views plus seven in-

tegrating subsystems. Next to the subsystem decomposition, the workflow of the metal printer is shown. This workflow is used to create a simple cycle-time model as formula. Note that, in the original A3, the formula was annotated with actual performance numbers to provide numerical insight into the cycle time.

At the top right-hand side of the A3, a customer key-driver graph is shown, and below the graph the key performance parameters are summarized.

This single A3 shows the system, the system context, and the first level of decomposition. Physical views and functional views are shown. Quantifications are given at all three levels as time-line, table, or formula.

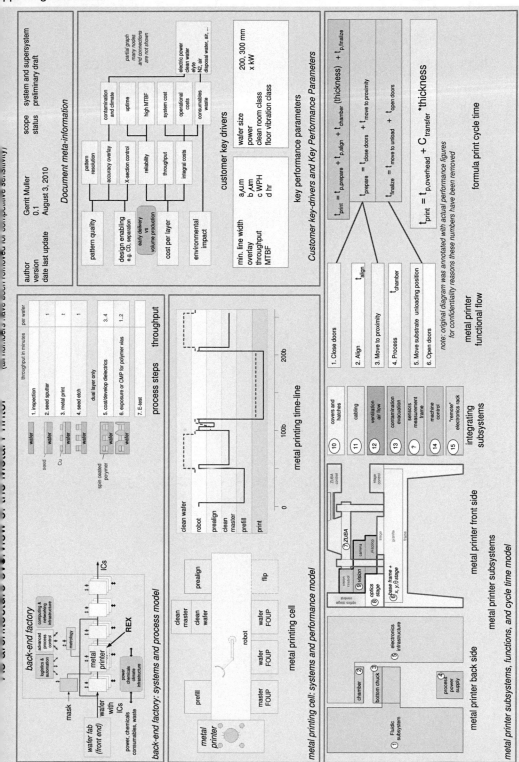

gure 7.7 Example of an Architecture Overview on one A3

EXERCISES

IN CLASSROOM FOR STUDENTS WITH WORKING EXPERIENCE

Create an overview of the process and the structure of the Product Creation documentation. Annotate strong aspects of process and structure and identify improvement opportunities. Present the results in one flipchart.

IN CLASSROOM FOR STUDENTS WITHOUT WORKING EXPERIENCE

Create a proposal for the structure of the Product Creation documentation, especially at the systems level. Pay special attention to the decomposition directions, such as construction, functional, and qualities. Provide the rationale for the chosen structure. Present the results in one flipchart.

8 Systems and Software

8.1 THE ROLE OF SOFTWARE IN SYSTEMS

8.1.1 INTRODUCTION

The relation between the software and systems disciplines is difficult in many organizations. The poor relation between the disciplines results in gaps in the design and later in quality problems in the final systems. As a consequence, software in many organizations is perceived as a problem and as a bottleneck in product creation.

Part of the explanation is that traditionally physical-oriented disciplines, such as mechanical, optical, or electrical engineering, dominated system design. Historically, engineers from these physical disciplines were confronted most with the application domain. These engineers have evolved into domain engineers.

Software has a significant impact on many system qualities in the modern world, as we will show in this chapter. More and more customer value depends on software. Unfortunately, many software engineers have not yet built up sufficient knowledge of the physical aspects of their systems or of application domain. At the same, time the engineers from the physical disciplines, who dominate the system design, do not yet understand the jargon and the concepts from the "virtual" disciplines (software, digital electronics engineering).

8.1.2 WHY IS SOFTWARE A BOTTLENECK IN DEVELOPMENT?

Growth of software effort

Software is a relative young discipline. The amount of software in systems is growing exponentially. The contribution of different disciplines to the system, measured in effort, is shifting continuously. Figure 8.1 shows the growth of the effort to make software and the related relative decrease in that of other disciplines.

Roles of disciplines in a system

The different disciplines do have an asymmetric relation when we look at the control in systems. Figure 8.2 shows a typical control hierarchy in a system. At the bottom we see the physical disciplines that realize physical devices and sensors. We prefer to keep these physical components independent of each other, seen from the control perspective. Safety provisions are the major exception to this rule; some first-line safety measures are realized in hardware.

Physical devices need actuation that is delivered by some analog (power) electronics, such as amplifiers. Note that there might be all kinds of conversions in between in the more complex reality, for example, pressure in a hydraulic system or light in an optics system. Again, we prefer to keep the analog electronics mutually independent. Analog electronics is controlled by digital electronics. The control stack continues

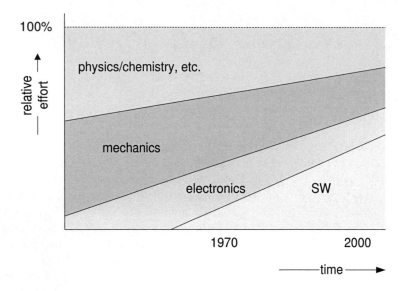

Figure 8.1 The relative contribution of software effort as a function of time.

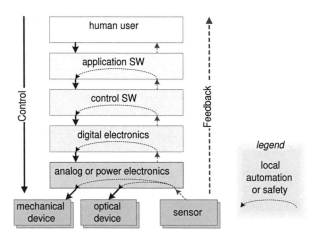

Figure 8.2 The control hierarchy of a system along the technology dimension.

with control software that sits on top of the digital hardware. Finally, application software determines what the control software should do. Hopefully, the human user is the person who is really in control.

Note that, in all layers, there are several reasons to have shortcuts from sensors to control:

Safety is always kept as simple and direct as possible since any complexity introduces new safety risks. A good safety design carefully allocates safety functions to the different layers to achieve the desired safety while achieving the desired control flexibility.

Automation can be done on lower layers if this simplifies the overall design. Automation provides value when the higher-level workflows are well understood and well defined.

Performance is a special case of automation where the shortcut facilitates better performance, for example, fast response times.

The software technology is, in most modern systems, the integrating technology, as shown by the control hierarchy. We will dive somewhat deeper into the relation between system qualities and software technology in the next section. Software technology determines, to a high degree, most system qualities in modern systems. The physical design often determines the inherent system qualities, but software construction often determines the actually achieved quality. For example, we can dimension a system with quite powerful motors to ensure high performance, but if the software does not fully utilize the motors, then the system performance is lower than can be expected from the physical design; similarly for reliability that inherently is determined by the physical design. However, software control may negatively impact reliability. For example, in a system with pumps, the software used a sequence where one of the pumps regularly ran dry. The consequence was that this pump often failed.

Characterization of disciplines.

Physical disciplines work on aspects that can be touched; the subjects are tangible. Virtual disciplines work on abstract concepts, and the subjects are intangible. Figure 8.3 shows the disciplines on an axis of decreasing tangibility and increasing abstractness. Mechanical engineering working with highly tangible constructions is one of the older disciplines. Analog (power) electronics is younger as a discipline and less tangible. Digital electronics is again younger. Although digital electronics itself can be touched, the circuitry itself is much more conceptual and abstract.

Figure 8.3 also provides a number of other characterizations that follow the same trend as tangibility and abstractness:

Maturity The more tangible, the more mature a discipline seems to be. Mature means here well known and founded; the discipline has an established and documented body of knowledge.

Production lead-time The physical world is constrained by nature. Processing and production of components have an inherent lead-time. Software can be seen as in-

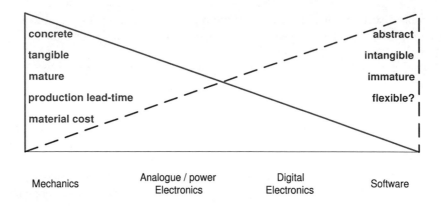

Figure 8.3   Characterization of disciplines, ordered along the level of abstraction

finitely fast. However, when testing, quality control and configuration management are included in the production lead-time, and then this lead-time becomes strongly dependent on people, processes, and tools. Hence the question mark behind flexible on the right-hand side of the figure.

Material cost Physical systems do have inherent cost in the materials and its processing.

These differences in nature, especially *production lead-time* and *material cost*, also cause differences in other business processes and the approach to life-cycle aspects. For many physical components, the logistics design is crucial for manufacturing cycle time, stocks, and cost, where software does have zero reproduction cycle time, cost, and infinite stocks.

8.1.3   SYSTEMS OR SOFTWARE ISSUES?

Systems can be specified in terms of their functionality and qualities. The software design influences or even determines most qualities of a system. Figure 8.4, based on [14], shows a checklist for qualities. In this figure, all qualities that have a strong or weak relation with the software design are highlighted.

The system is decomposed in subsystems and implementation technologies during Systems Design. The combination of subsystems and technologies together has to realize the qualities. During this step, the contribution or role of a subsystem and implementing technology is determined.

Figure 8.5 shows the system-level design aspects that are strongly related to software. Figure 8.6 shows a list of mechanisms used by software engineers. These mechanisms facilitate the system-level design aspects mentioned in Figure 8.5.

Both *Quality Attributes* and *Design Aspects* are *system-level* issues; however, the software predominantly influences most of these issues. The Systems Architect should

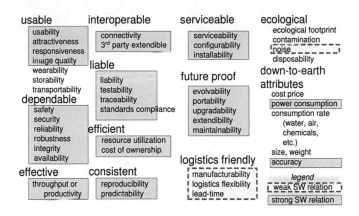

Figure 8.4   Quality checklist annotated with the relation with software.

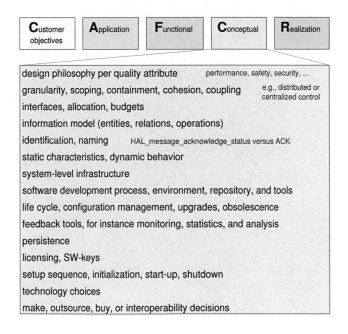

Figure 8.5   System design aspects that are strongly SW related.

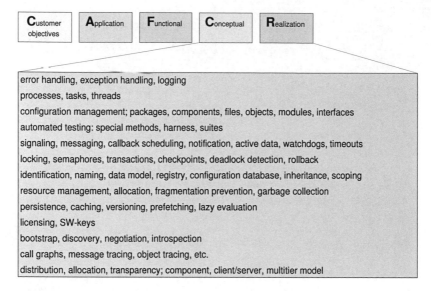

Figure 8.6 List of software mechanisms that are frequently applied to solve system-level design aspects.

- Define the system level **what**.
- Codesign the system level **how**.
- Be involved with the single technology or subsystem **how**.

    Due to the strong software impact, the Software Architect should

- Understand and review the system-level **what**.
- Codesign the system-level **how**.
- Design the software **how**.

This requires significant domain expertise of the software architect.

    Figures 8.5 and 8.6 contain too many design aspects and software mechanisms to discuss as part of this book. The main purpose of these lists is to show the variety of technology issues to be addressed by the software architect.

    Many of the design aspects have a many-to-many relation to software mechanisms. For example, the design strategies for *performance*, *safety*, and *security* relate to nearly all software mechanisms. Vice versa, most software mechanisms penetrate throughout most software and relate back to most of the design aspects.

    The software part of systems is complex in itself. Software is a construct made by many people, stacking construct on construct. The risk is that software architects spend all their time internally in the software, while they also have to relate the software choices to the context, the system.

8.2 SYSTEM INTEGRATION: HOW TO

8.2.1 INTRODUCTION

Quality problems and delays are some of the symptoms of the troublesome relation between software and system. The integration of software and hardware is, in many organizations, taking place when both are nearly finished. Organizational boundaries propagate into the schedule, causing too late integration of crucial technologies. Systems architects have to ensure that software–hardware integration starts very early.

Systems Integration is one of the activities of the Product Creation Process. The Product Creation Process starts with a set of product needs and ideas, and results in a system that

- fits in customers' needs and context.
- can be ordered, manufactured, installed, maintained, serviced, and disposed.
- fits the business needs.

During product creation, many activities are performed, such as feasibility studies, requirements capturing, design, engineering, contracting with suppliers, verification, testing, etc. Decomposition is a universal method used in organization, documentation, and design. Decomposition enables the distribution of work in a concurrent fashion. The complement to decomposition is integration. Every activity that has been decomposed in smaller steps will have to be integrated again to obtain the single desired outcome.

Integration is an ongoing flow of activities during the entire product creation cycle. The nature of integration activities, however, shifts over time. Early on in the project, technologies or components are integrated, while at the end of the project the entire system is built and verified. In formal process descriptions,[1] the description of product integration is mostly limited to the very last phase of the total integration flow, with a focus on the administrative and process aspects. We use the term integration in the broader meaning of all activities where decomposed parts are brought together.

In practice, projects hit many problems that are caused by decomposition steps. Whenever an activity is decomposed, the decomposed activities normally run well; however, crosscutting functionality and qualities suffer from the decomposition. Lack of ownership, lack of attention, and lack of communication across organizational boundaries are root causes of these problems. The countermeasure for these problems is to have continuous attention on integration.

Goal of Integration

The goal of integration is to find unforeseen problems as early as possible in order to solve these problems in time. Integration plays a major role in risk reduction. The

---

[1] for example, NASA Procedural Requirements (NPR).

word "unforeseen" indicates the main challenge of integration: how to find problems when you do not know that they exist at all?

Problems can be unforeseen because the knowledge of the creation team is limited. May be nobody on earth did have the knowledge to foresee such a problem simply because the creation process enters new areas of knowledge. Problems can also be unforeseen due to invalid assumptions. For instance, many assumptions are being made early in the design to cope with many uncertainties. The limited intellectual capabilities of us, humans, limit also the degree into which we can oversee all consequences of uncertainties and the assumptions we make. A common source of unforeseen problems is interference between functions or components. For example, two software functions running on the same processor may perform well individually, but running concurrently may be excessively slow due to cache pollution or memory trashing.

Product Integration as part of the Product Creation Process

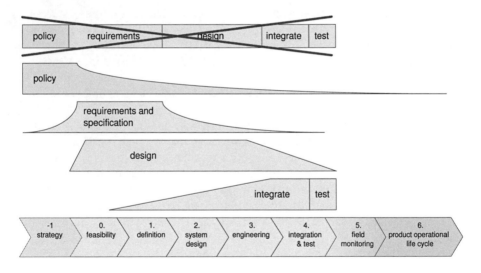

Figure 8.7 Typical Product Creation Process and the concurrency of engineering activities.

The Product Creation Process (PCP) is often prescribed as a sequence of phases with increasing level of realization and decreasing level of risk. This is a useful high-level mental model; however, one should realize that most activities have much more overlap in the current dynamic world. The pure waterfall model, where requirements, design, integration, and test are sequential phases, is not practical anymore. Much more practical is an approach with a shifting emphasis as shown in Figure 8.7. A comparable approach is Rational Unified Process (RUP); see [21] for this. and [22] for integration. Note especially the long ramp-up of the integration, the focus of this section.

Integration in Relation to Testing

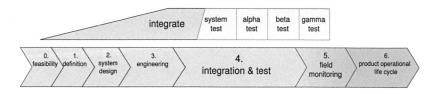

Figure 8.8 Zooming in on integration and tests.

Integration and testing are often used as identical activities. However, the two are related and completely different at the same time.

Figure 8.8 zooms in on the integration and test activities. Integration is the activity where we try to find the unknowns and where we resolve the uncertainties. Testing is an activity where we operate a (part of a) system in a predefined manner and verify the behavior of the system. A test passes if the result fits the pecified behavior and performance, and otherwise, it fails. Components are tested at component level before integration starts. Many tests may be applied as part of the integration. These tests are applied during integration to find these unknowns and to resolve the uncertainties. When the milestone, that the system is perceived to be ready is passed, then the systems engineers will run an entire system-level test suite. Normally, this run still reveals unknowns and problems. The system test verifies both the external specification as well as the internal design. When sufficient stability of the system test is achieved, a different working attitude is taken: from problem solving to verification and finishing. The alpha test starts with a hard milestone and is also finished at a well-defined moment in time. The alpha test is the formal test performed by the product creation team itself, where the specification is verified. The beta test is also a well-defined time-limited formal test performed by the "consuming" internal stakeholders: marketing, application, production, logistics, and service. It also verifies the specification, but the testers have not been involved in product creation. These testers are not blinded by their a priori knowledge. Finally, the external stakeholders, such as actual users, test the product. Normally, problems are still found and solved during these tests, violating the assumption that the system is stable and unchanged during testing. In fact, these alpha, beta, and gamma testers hit problems that should have been found during integration. We will focus the rest of this chapter on integration, with the main purpose being to reduce risks in the testing phase by identifying (potential) problems as early as possible.

8.2.2 WHAT, HOW, WHEN, AND WHO OF INTEGRATION

By necessity, the integration of a system starts bottom-up with testing individual components in a provisional component context. The purpose of the bottom-up steps is to find problems in a sufficiently small scope; the scope must be small enough to allow diagnosis in case of failure. If we bring thousands of components together into a system, then this system will fail for certain. But it is nearly impossible to find the sources of this failure due to the multitude of unknowns, uncertainties, and ill-functioning parts.

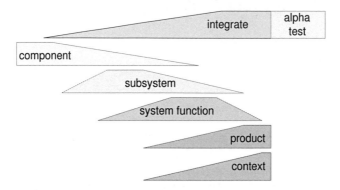

Figure 8.9   Integration takes place in a bottom-up fashion, with large overlaps between integration levels.

The focus of the integration activity is shifting during the integration phase. Figure 8.9 shows the bottom-up levels of integration over time. Essential to integration is that the higher levels of integration already start when the lower levels of integration are not yet finished. The different levels of integration are therefore overlapping. Early during integration, the focus is on functionality and behavior of components and subsystems. Then the focus shifts to system-level functionality: do the subsystems together operate in the right way? The last step in integration is to focus on system qualities, such as performance and reliability.

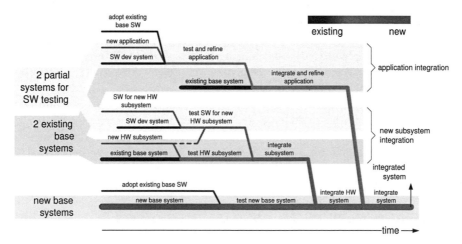

Figure 8.10   During integration, a transition takes place from using previous systems and partial systems to the new system configuration.

The integrator tries to integrate subsystems or functions as early as possible with the purpose of finding unforeseen problems as early as possible. This means that integration already takes place, while most of the new components, subsystems, and functions are still being developed. Normally, partial systems or modified existing systems are used in the early phases of integration as a substitute for the not-yet-available parts. Figure 8.10 shows this transition from using partial and existing subsystems to systems based on new developed parts.

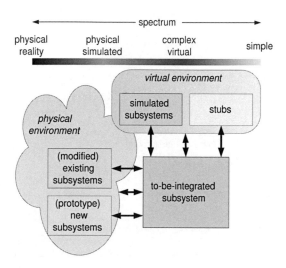

Figure 8.11   Alternatives to integrate a subsystem early in the project.

The unavailability of subsystems or the lack of stability of new subsystems forces the integrator to use alternatives. Figure 8.11 shows a classification of alternatives. Simple stubs in a virtual environment up to real physical subsystems in a physical environment can be used. In practice, multiple alternatives are combined. As a function of time, the integration shifts from the use of stubs and a virtual environment to a scenario as close as possible to the final physical reality.

The challenge for the project team is to determine what intermediate integration configurations are beneficial. Every additional configuration adds costs: creation costs as well as costs to keep it up to date and running. An even more difficult conflict is that the same critical resources, for instance, dynamic positioning experts, are needed for the different configurations. Do we focus completely on the final product, or do we invest in intermediate steps? Final yet important is the configuration management problem that is created with all integration configurations. When hundreds or thousands of engineers are working on a product, then most of them are in fact busy with changing implementations. Strict change procedures for integration configurations may reduce the management problem, but this conflicts often with troubleshooting needs during integration.

Crucial questions in determining what intermediate configurations to create are

• How critical or sensitive is the subsystem or function to be integrated?

- What are the aspects that are sufficiently close to final operation so that the feedback from the configuration makes sense?
- How much needs to be invested in this intermediate step? Special attention is required for critical resources.
- Can we formulate the goal of this integration system in such a way that it guides the configuration management problem?

1	Determine most critical system performance parameters.
2	Identify subsystems and functions involved in these parameters.
3	Work towards integration configurations along these chains of subsystems and functions.
4	Show system performance parameter as early as possible; start with showing "typical" system performance.
5	Show "worst-case" and "boundary" system performance.
6	Rework manual integration tests in steps into automated regression tests.
7	Monitor regression results with human-driven analysis.
8	Integrate the chains: show system performance of different parameters simultaneously on the same system.

Figure 8.12 Stepwise integration approach.

Based on these considerations, we propose a stepwise integration approach as shown in Figure 8.12. The first step is to determine a limited set of the most critical system performance parameters such as image quality, productivity, or power consumption. These system performance parameters are the outcome of a complicated interaction of system functions and subsystems; we call the set of functions and subsystems that result in a system parameter a chain. We start to define partial system configurations as integration vehicles once we have identified critical chains. The critical chains serve as guidance for the integration process.

We strongly recommend focusing on showing the critical system performance parameters as early as possible. The focus is on "typical" performance in the beginning. We get room to study "worst-case" and "boundary" performance once the system gets somewhat more stable and predictable.

It is important to monitor the system performance regularly since many engineers are still changing many parts of the total system. The early integration tests are manual tests because the system circumstances are still very premature and because integrators have to be responsive to many unexpected problems. In due time the chain and the surrounding system get more stable, allowing automation of tests. We can migrate the early manual integration steps into an automated regression test.

The results of regularly performed regression tests must be monitored and analyzed by system engineers. This analysis does not focus on pass or fail, but rather, looks for trends, unexplained discontinuities, or variations.

Later during integration, we have to integrate the chains themselves and show the simultaneous performance of the critical performance parameters.

The foregoing approach requires quite some logistics support. The project leader will therefore make integration schedules in close cooperation with systems engineers. Integration schedules have two conflicting attributes:

Predictability and stability to ensure timely availability of resources
Flexibility and agility to cope with the inherent uncertainties and unknowns

The starting point to create a schedule is to determine a specific and detailed integration order of components and functions. The integration order is designed such that the desired critical system performance parameter can be measured as early as possible.

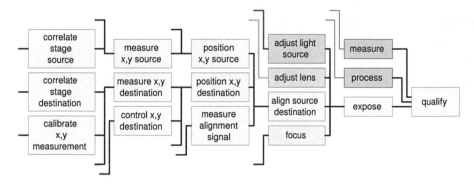

Figure 8.13 Example of small part of the order of functions required for the image quality system performance parameter of a wafer stepper.

Figure 8.13 shows an example of a specific order of functions required to determine the image quality system performance parameter of a wafer stepper. Such a diagram starts often on the right-hand side: what is the desired output parameter to be achieved? Next, the question "What is needed to achieve this output?" is asked recursively. This very partial diagram is still highly simplified. In reality, many of these functions have multiple dependencies.

A worse case is that circular dependency often exists. For instance, in order to align source and destination, we need to be in focus, while in order to find the focal plane, we need to be aligned. These dependencies are known during design time and already solved at that moment. For example, a frequently used design pattern is a stepwise refined one: coarse and fine alignment, and coarse and fine focusing. The creation of a detailed integration schedule provides worthwhile inputs for the design itself. Making the integration schedule specific forces the design team to analyze the design from the integration perspective, and it often results in the discovery of many (unresolved) implicit assumptions.

The existence of this integration schedule must be taken with a grain of salt. It has a large value for the design and for understanding the integration. Unfortunately, the integration process itself turns out to be poorly predictable: it is an ongoing set of crises and disruptive events, such as late deliveries, breaking-down components, non-functioning configurations, missing expertise, wrong tolerances, interfering artifacts, etc. Crucial to the integration process are capabilities to improvise and troubleshoot.

The integration schedule is a rather volatile and dynamic entity. It does not make sense to formalize the integration heavily, neither to keep it updated in all details. Formalization and extensive updating takes a lot of effort with little or no bene-fits. The recommended approach is to use the original integration schedule as kind of reference and to use short cyclic planning steps to guide the integration process. Typ-ical meeting frequency during integration is once per day. Every meeting results and problems, required activities and resources, and short-term schedule are discussed.

During integration, many project team members are involved with different roles and responsibilities:

- Project leader
- Systems architect/engineer/integrator
- System tester
- Logistics and administrative support personnel
- Engineers
- Machine owner

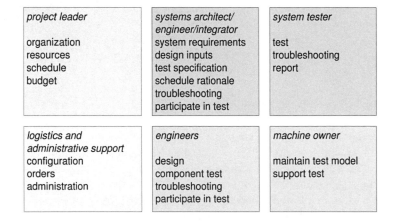

Figure 8.14   Roles and responsibilities during the integration process.

Figure 8.14 shows these roles in relation to their responsibilities. Note that the actual names of these roles depend on the organization; we will use these generic labels in this section.

The *project leader* is the organizer who takes care of managing resources, schedule, and budget. Based on inputs from the systems engineer, the project leader will claim and chase the required resources. The project leader facilitates the integration process. This contribution is critical to project timing.

The *systems architect*, *systems engineer*, and *systems integrator* roles are in fact a spectrum of roles that can be performed by one or more persons, depending on their capabilities. A good systems architect is sometimes a bad systems integrator and vice versa[2]. This role is driven by content, relating critical system performance parameters to design and test. In this role, the rationale of the integration schedule is determined, it being a joint effort of the project leader and systems engineer. The integral perspective of this role results in a natural contribution to troubleshooting.

The *systems tester* is the practical person actually performing most of the tests. During the integration phase, much of the time of the systems tester is spent in troubleshooting, often of trivial problems. More difficult problems are escalated to engineers or the systems integrator. The system tester documents test results in reports.

The *machine owner* is responsible for maintaining a working, up-to-date, test model. In practice, this is quite a challenging job because many engineers are busy with doing updates and performing local tests, while the systems integrator and systems tester need undisturbed access to a stable test model. We have observed that explicit ownership of one test model by one machine owner increases the test model stability significantly. Organizations without such a role lose a lot of time due to test model configuration problems.

Engineers deliver locally tested and working components, functions, or subsystems. owever, the responsibility of the engineers continues into the integration effort. They participate in integration tests and help in troubleshooting.

The project team is supported by many kinds of support personnel. For integration, *logistics and administrative support* is crucial. Logistics suport people perform configuration management of test models as well as the products to be manufactured. Note that integration problems may induce changes in formalized product documentation and logistics of the final manufacturing, which can have significant financial consequences due to the concurrency of development and preparation for production. The logistics support people also have to manage and organize unexpected but critical orders for parts of test models.

8.2.3 CONFIGURATION MANAGEMENT

Configuration management and integration are intimately related as discussed in previous sections. We should realize that configuration management plays a role in many processes. Figure 8.15 shows a simplified process decomposition of those processes elated to configuration management.

---

[2]Critical characteristics for architects are the balance between theoretical versus hands-on, conceptual versus implementation, and creative and diverging versus result-driven and converging. The emphasis must be on hands-on, implementation, and result driven and converging during integration.

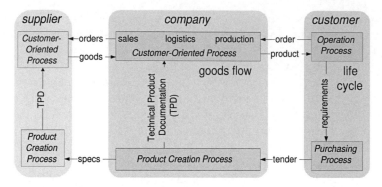

Figure 8.15 Simplified process diagram that shows processes that are relevant from the configuration management perspective.

Basically, the internal *Customer-Oriented* and *Product Creation* processes are linked to the related supplier and customer processes. There are two main flows where configuration management plays a role:

- Creation flow, from customer requirements to component specifications to technical product documentation to be used in the other flow
- Goods flow, a repeating set of processes where orders are fulfilled by a logistics and production chain

In principle, the creation flow is an one-time project activity. This flow may be repeated to create successor products, but this is a new instantiation of this flow. The goods flow is a continuous flow with life-cycle considerations. The final product as used operationally by customers also has its own life cycle.

Many entities have changing configurations and therefore need configuration management. Figure 8.16 shows the same process decomposition as in Figure 8.15, but now annotated by entities under configuration management. Two classes of configuration management entities exist: information and physical items. The information entities are normally managed by procedures and computer-based tools. However, for physical entities, the challenge is to maintain consistency between the actual physical item and the data in the configuration management administration. Especially during the hectic period of integration, the administration sometimes differs from the physical reality, causing many nasty problems. Sometimes more effort in processes helps; however, this results in more latency and more work-around behavior. Unfortunately, there is no silver bullet for configuration management processes.

The main configuration management entities during integration are the test models. Changes in test models may have to propagate to other entities, such as specifications, technical product documentation, and, due to concurrency, to components and products in the goods flow processes.

One particular area of attention is the synchronization of components, subsystems, and test models. All these entities exist and change concurrently. A certain pull to

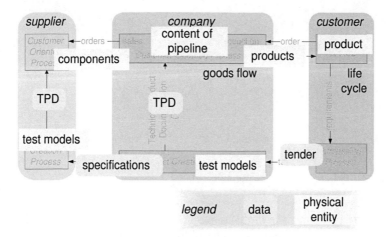

Figure 8.16 The simplified process diagram annotated with entities that are under configuration management.

use latest versions is caused by the fact that most problems are solved in the latest version. However, integrators and testers need a certain stability of a test model. This makes them hesitant to take over changes. One should realize that only a limited amount of test models exist, while all these engineers create thousands of changes. On top of this problem comes a logistics problem: from change idea to availability of changed component or function may take days or weeks. Sometimes one single provisionally changed component is available early.

One way of coping with the diversity of test model configurations is to formulate clearly the integration goals of the different test models. Note that these integration goals may change over time, according to Figure 8.9.

8.2.4 TYPICAL ORDER OF INTEGRATION PROBLEMS OCCURRING IN REAL LIFE

Experience in many integration phases resulted in the observation of a typical order when integration problems occur. This typical order is shown in Figure 8.17.

Typically, none of these problems should occur, but despite mature processes, all of them occur in practice. The failure to build the system at all is often caused by the use of implicit knowledge. An example is a relatively addressed data file that resides on the engineer's workbench but that is not present in the independent test environment. As a side remark, we observe the tension between using networked test models. Network connections shorten software change cycles and help in troubleshooting, however; at the same time, the type of problems we discussed here may stay invisible.

The next phase in integration appears to be that individual components or functions work but cease to function when they are combined. Again, the source of the problem is often a violated implicit assumption. This might relate to the third

1.	The (sub)system does not build.
2.	The (sub)system does not function.
3.	Interface errors.
4.	The (sub)system is too slow.
5.	Problems with the main performance parameter, such as image quality.
6.	The (sub)system is not reliable.

Figure 8.17   Typical order of integration problems

problem, interface errors. The problem might be in the interface itself; for instance, different interpretations of the interface specification may result in failures of the combination. Another type of problem in this category is again caused by implicit assumptions. For example, the implementation of the calling subsystem is based on assumed functionality of the called subsystem. It will be clear that behavior that is different from what is assumed for the called subsystem may cause problems for the caller. These types of problems are often not visible at the interface specification level because none of the subsystem designers may have realized that the behavior is relevant at interface level.

Once the system gets operational functionally, then the nonfunctional system properties become observable. The first problem that hits integrators in this phase is often system performance in terms of speed or throughput. Individual functions and components in isolation perform well, but when all functionality is running concurrently sharing the computing resources, then the actual performance can be measured. The mismatch of expected and actual performance is not only caused by concurrency and sharing but also by the increased load of more realistic test data. On top of these problems, nonlinear effects appear when the system resources are more heavily loaded, worsening overall performance even more. After some redesigns, the performance problems tend to be solved, although continuous monitoring is recommended. Performance tends to degrade further during integration due to added functionality and solutions for other integration problems.

When the system is both functional and well performing, then the core functionality, the main purpose of the product, is tested extensively. The application experts are closely involved in integration in this phase. These application experts use the system differently and look differently at the results. Problems in the critical system functionality are discovered in this phase. Although these problems were already present in the earlier phases, they stayed invisible due to the dominance of the other integration problems and due to the different perspectives of technical testers and application experts.

During the last integration phase, the system gets used more and more intensively. The result is that less robust parts of the design are exercised more, causing system crashes. A common complaint in this phase is that the system is unreliable and unstable. Part of this problem is caused by the continuous influx of design changes triggered by the earlier design phases; every change also triggers new problems.

EXERCISES

IN CLASSROOM FOR STUDENTS WITH WORKING EXPERIENCE

Create a simple functional model of the software in your system. Identify the key software technologies in your system, including operating systems, and programming languages. Make a presentation with functional model and key technologies on a flipchart where potential problem areas are annotated.

IN CLASSROOM FOR STUDENTS WITHOUT WORKING EXPERIENCE

Perform the following tasks for the product provided by your teacher:

1. Create a simple functional model of the software in your system.
2. Identify the potential key software technologies in your system, including operating systems, and programming languages.
3. Identify the most critical software design decisions.

9 Boardroom Presentation

9.1 INTERMEZZO: ARCHITECT VERSUS MANAGER; THE TENSE RELATION

9.1.1 INTRODUCTION

The relation between managers and systems architects somehow tends to be somewhat difficult. This is not desired since we position the systems architects as part of the leadership of an organization.

In this intermezzo we look at managers and architects in a generalized way. Generalizations are always risky; the purpose of this generalization is to better understand the inherent tensions between architects and managers. Neither "real" architect nor manager will exactly look like the generalization in this intermezzo.

9.1.2 WHAT IS A MANAGER?

A manager is someone who manages everything needed to get a task executed. The manager has the responsibility for the task. The responsibility comes with the required authority to do the task. Every process in the simplified business decomposition in Chapter 1, Sections 1.1 and 1.3, generally has a manager associated with the process who is responsible for the execution of that process. Often these tasks are further decomposed with managers associated with every subtask.

Systems architects frequently encounter the managers shown in Figure 9.1.

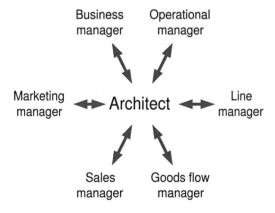

Figure 9.1 Managers frequently interacting with architects.

9.1.3   COMPARISON OF ARCHITECT AND MANAGER

Figure 9.2 shows a comparison between architects and managers for six different aspects: responsibility, view on solutions, view on changes, personal characteristics, leadership values, *and* personal ambition.

responsibilities	architect	manager
scope	wide	limited
formal weight	low	high

view on solutions	architect	manager
design	elegant	if it works it is OK
application	perfect fit	no complaints
future proof	important	task dependent

view on changes	architect	manager
viewpoint	changes needed: + stakeholders + time + problem analysis	changes introduce: – problems – uncertainties – new changes
attitude	fact of life	avoid changes

	architect	manager
personal characteristics	independent critical curious	conformance demanding control minded

leadership values	

based on
  knowledge
  vision
	based on
	  KPI's
	title creates
	  expectations
	task driven

personal ambition	
best solutions	highest hierarchical level

Figure 9.2   Comparison of caricature of architect and manager.

Responsibility

Managers have a well-defined responsibility, related to their function. They also are empowered accordingly in most organizations. The scope of responsibility is limited; the total responsibility is divided over many managers.

The responsibility of the architect is much fuzzier; see Chapter 2, Section 2.3. For every aspect the architect is working on, there is a manager who has the formal responsibility for that specific subject. The architect has limited formal power. At the other hand, architects have a lot of informal influence.

View on Solutions

The view on solutions is quite different. The architect partially trusts his or her intuition, looking for the notion of an elegant solution. The word elegant can cover many aspects, such as balanced, simple, beautiful. As representative of the stakeholders, the architect will ensure that the solution fits the needs: is it the "right" solution? At the same time, the architect will place the solution in a time perspective: is the solution "future proof"?

Most managers stay close to their task and responsibility. A solution that matches the specification is, by definition, good. If there are no complaints, then there is no problem.

View on Changes

Architects (ought to) spend a significant part of their time in the turbulent outside world, inhabited by demanding customers in changing markets with aggressive inventive competitors, and innovative suppliers. At the same time, architects are active in the company across many internal boundaries, enabling them to detect, analyze, and to help solve many internal problems. Architects are continuously confronted with situations where change is required. The internal and external worlds are highly dynamic, causing need for change everywhere. Architects see changes as a fact of life.

Managers tend to take an opposite view on the need for change caused by the limited scope and the heavy weight of the responsibility for the results of their tasks. Managers have experienced that changes always introduce problems, involve uncertainties, and trigger more changes. The resulting behavior is to avoid changes.[1]

Personal Characteristics

Managers are control-minded and like to be in control of the task being performed; that is exactly their job. Managers demand conformance as a means to be in control. The people working at a task have to conform to the way the manager wants to have the task performed.

Architects have an entirely different personality. They need independence and curiosity to be able to act as a representative of the stakeholders. At the same time, architects need to be critical: is the chosen approach the best way to do address the task?

Leadership Values

Many organizations still think in hierarchical terms. Hence the manager is seen as the person who sets the direction. However, it is questionable if managers do have the appropriate knowledge and vision to determine the direction.

Architects have a broad perspective and knowledge, while (good) architects also have vision. This is a natural combination for providing true leadership.

Some architects are handicapped by an introvert personality, making it difficult to "sell" the vision and to take the leader's position. It will be clear that the teamwork of manager and architect will work wonders in such a case.

---

[1] Keep being aware that we are discussing caricatures of architects and managers. In practice, there are many (bad) architects behaving very conservatively.

Personal Ambition

The personal ambition of managers and architects are opposite as well. Many managers are driven by normal career incentives: higher position, power, status, and more money. Architects seem to be driven by the case at hand; they want to achieve the "best" solution.

This difference in ambition makes architects difficult to control because they are rather insensitive toward normal incentives such as promotions and salary raises.

9.1.4   HOW TO IMPROVE THE RELATIONSHIP

The starting point for any solution is the recognition of the problem. This intermezzo is primarily provided to create awareness of the problem that there is tension between architects and managers. No silver bullets are given here as solution.

A quite promising direction in which to address this problem is that modern management techniques; see Figure 9.3 for a list of suggestions.

Empowerment

Delegation

Leadership instead of task-driven management

Process orientation instead of hierarchical organizations

Teamwork

Mutual Respect

Recognition of diversity and nonconformity

Reverse Appraisal

Stimulating open communication

Figure 9.3   List of modern management techniques that can be used to improve the relation between managers and architects.

Architects play a vital role in bootstrapping these management techniques. They play the role of a catalyst in many techniques due to the combination of personal characteristics such as independence and knowledge. If architects hide in technological solutions, then they do not trigger the required change.

We can also work on both sides to improve this relationship. Architects can be stuck in the solution world with little attention to all the nontechnical aspects that determine the architecture. A vital step is that they learn to communicate better what the impact of technical choices is on the less technical business aspects. Once architects are able to communicate more clearly with managers, then their recognition and influence will increase. See the next chapter for further elaboration.

Many managers do not know what to expect from architects. It helps managers if they do understand the role of the architects better, so that they can ask the right questions and provide coaching. This book can be used in courses directed to management teams to help them understand the architecting role.

9.2 HOW TO PRESENT ARCHITECTURE ISSUES TO HIGHER MANAGEMENT

9.2.1 INTRODUCTION

The architect bridges the technology world with other business related worlds by understanding these other worlds and by having ample knowledge of technologies. Management teams are responsible for the overall business performance, which in the end is expressed in financial results.

Many architects and management teams are caught in a vicious circle:

- Architects complain about management decisions and lack of expertise of managers.
- Managers complain about lack of input data and invisible architects.

One way to break this vicious circle is to improve the managerial communication skills of architects. We address a frequently needed skill: presenting an architecture issue to a management team.

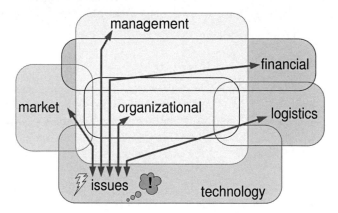

Figure 9.4 Architectural issues related to managerial viewpoints.

The architect should contribute to the managerial decision process by communicating technology options and consequences of technological decisions. Figure 9.4 shows a number of the relevant, somewhat overlapping, viewpoints. The figure indicates what links architects should communicate to management teams.

Architects must have a good understanding of their target audience. Figure 9.5 characterizes the managers in management teams. Their main job is to run a healthy business, which explains many of these characterizations: *action-oriented, solution rather than problem, impatient, busy, bottom-line oriented*, and *want facts not believes*. Bottom-line orientation means focus on profit, return on investment, market *common characteristics*

+ action-oriented

+ solution rather than problem

+ impatient, busy

+ want facts not beliefs

+ operate in a political context

+ bottom-line oriented:
   profit, return on investment,
   market share, etc.

highly variable characteristics

? technology knowledge
   from extensive to shallow

? style from power play to
   inspirational leadership

Figure 9.5   Characteristics of managers in higher-management teams.

share, etc. These managers operate with many people all with their own personal interests. This means that managers *operate in a political context* (something that architects like to ignore).

Some characteristics of management teams depend on the company culture. For example, the amount of technology knowledge can vary from extensive to shallow. Or, for example, the management style can vary from power play to inspirational leadership.

9.2.2   PREPARATION

Presentations to higher-management teams must always be prepared with multiple people: a small preparation team. The combined insights of the preparation team enlarge the coverage of important issues, both technical as well as business. The combined understanding of the target audience is also quite valuable. Figure 9.6 shows how to prepare the content of the presentation as well as how to prepare for the audience.

The content of the presentation must be clear, address the main issues, and convey the message; see also Section 9.2.3. The message must have substance for managers, which means that it should be *fact based*. The first steps are *gathering facts* and *performing analysis*. Based on these facts, the *goal* and *message* of the presentation must be articulated. All this information must be combined in a *presentation*. When the presentation content is satisfactory, the form must be polished (templates, colors, readability, etc.). Although this has been described as a sequential process, the normal incremental spiral approach should be followed, going through these steps in two to three iterations.

The members of management teams operate normally in a highly political context, mutually as well as with people in their context. This politics interferes significantly with the decision making. The preparation team should map the political situation; the team must identify and understand the political forces. This is done by *analyzing*

Always prepare with small team!

content	mutual interaction	understand audience
70% of effort		30% of effort
+ gather facts		
		+ gather audience background
+ perform analysis		
		+ analysis audience interests
+ identify goal and message		
		+ identify expected responses
+ make presentation		
		+ simulate audience, exercise presentation
+ polish presentation form		

Figure 9.6   How to prepare.

the audience, their *background*, and their *interests*. The preparation team can gain a lot of insight by discussing the *expected responses* of the management team. At an appropriate moment, the preparation team can *simulate* (role-play) the management team in a proof-run of the presentation. The understanding of the audience must be used to select and structure the content part of the presentation. This activity should be time-multiplexed with the content preparation; 70% of the time working on content, 30% of the time for reflection and understanding of the audience.

9.2.3   THE PRESENTATION MATERIAL

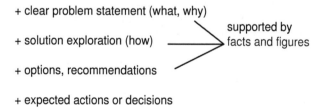

+ clear problem statement (what, why)

+ solution exploration (how)   supported by
                                       facts and figures
+ options, recommendations

+ expected actions or decisions

Figure 9.7   Recommended content.

Figure 9.7 provides guidelines for the contents of the presentation. A clear *problem statement* and an *exploration of solutions* should address the technical issues as well as the translation to the business consequences. Normally, a range of options is *provided*. The options are *compared*, and *recommendations* are provided. Note that options that are unfavorable from architectural point of view are nevertheless options. It is a challenge for the architect to articulate why these options are bad and should not be chosen. Architects enable and streamline the decision making by providing clear recommendations and by indicating what *actions* or *decisions* are required.

All content of the presentation should be to the point, *factual*, and *quantified*. Quantified does not mean certain, often quite the opposite; future numbers are estimates based on many assumptions. The reliability of the information should be evident in the presentation. Many facts can be derived from the past. *Figures* from the past are useful to "calibrate" future options. Deviations from trends in the past are suspect and should be explained.

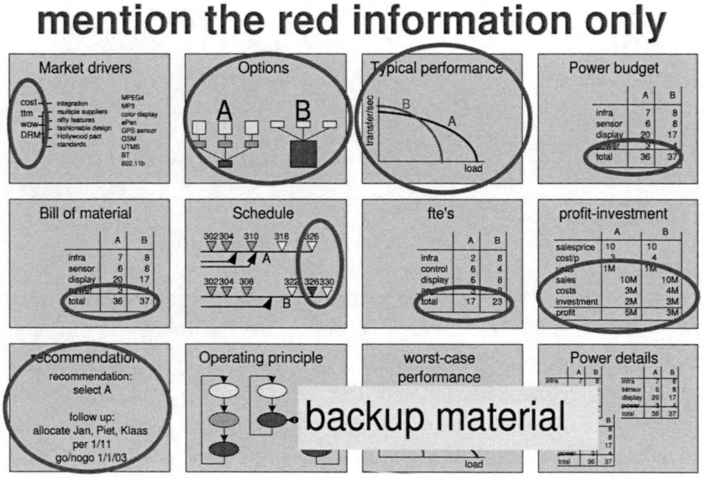

Figure 9.8 Mentioned info, shown info and backup info.

The presentation material should cover more than is actually being presented during the presentation itself. Some supporting data should be present on the sheets without mentioning the data explicitly during the presentation. This allows the audience to assess the validity of the presented numbers, without the need to zoom in on all the details.

It is also useful to have additional backup material available with more in-depth supporting data. This can be used to answer questions or to focus the discussion: speculation can be prevented by providing actual data (Figure 9.8.

The use of demonstrators and the show of artifacts (components, mockups) make the presentation livelier. The demonstrations should be short and attractive (from the customer point of view), while illustrating the value and technological possibilities and issues.

Architects prefer to focus on the content; form is supportive to of the transfer of the content. However, architects should be aware that managers can be distracted by the form of a presentation, potentially spoiling the entire meeting by small issues. Figure 9.9 gives a number of recommendations with respect to the form of the presentation and the appearance of the presenter.

The presentation material (slides, demonstrators, video, drawings, etc.) have to poor form can easily distract from purpose and content

presentation material	presenter's appearance
+ professional	+ well dressed
+ moderate use of color and animations	+ self confident but open
+ readable	
+ use demos and show artifacts	but stay yourself, stay authentic

Figure 9.9 Form is important.

look professional. Slides will use color and other presentation features. However, moderation in the use of color, animation and other presentation features is recommended; an overload of these does not look professional, and will distract the audience from the actual content. Information on the slides has to be readable: use large-enough fonts and use sufficient contrast with the background. Pay special attention to quality and readability; especially when copy-pasting information from other sources the quality of presentation is often degraded. Sometimes it is better to recreate a high-quality table or graph than to save effort by copy-pasting an unreadable table or graph.

The appearance of the presenter can also make or break the presentation. The presenter should give sufficient attention to clothes and overall appearance. Do not exaggerate this,; you should stay yourself and still be authentic. Other people immediately sense it when the appearance is too exaggerated, which is also damaging to your image.

9.2.4 THE PRESENTATION

do not	do
- preach beliefs	+ quantify, show figures and facts
- underestimate technology knowledge of managers	+ create faith in your knowledge
- tell them what they did wrong	+ focus on objectives
- oversell	+ manage expectations

Figure 9.10 Do not force your opinion; understand the audience.

Figures 9.10 and 9.11 show in the *do not* column a number of pitfalls for an architect when presenting to higher-management teams. The preferred interaction pattern is given in the *do* column.

The pitfalls in Figure 9.10, *preaching beliefs*, *underestimating expertise of managers*, and *telling managers what they did wrong*, are often caused by insufficient understanding of the target audience. In these cases, the opinion of the architect is too dominant; opinions work in counterproductive ways. *Overselling* creates a problem for the future: expectations are created that cannot be met. The consequence of overselling is loss of credibility and potentially lack of support in tougher times. Architects must *manage* the *expectations* of the audience.

When presenting, the architect tries to achieve multiple objectives:

- Create awareness of the problem and potential solutions by *quantification* and by *showing figures and facts*.
- Show architecting competence in these areas, with the message being "You, the manager, can delegate the technical responsibility to me." This creates *faith* in the *architect's knowledge*.
- Facilitate decision making by translating the problem and solutions in business consequences, with the *focus on objectives*.

This means that sufficient technological content need to be shown, at least to create faith in the architect's competence. Underestimation of managerial knowledge is certainly arrogant, but mostly very dangerous. Some managers have a significant historic knowledge, which enable them to assess strengths and weaknesses quickly. Providing sufficient depth to this type of manager is rewarding. The less-informed manager does not need to fully understand the technical part but at least should get the feeling that he or she understands the issues.

do not	*do*
- let one of the managers hijack the meeting	+ maintain the lead
- build up tensions by withholding facts or solutions	+ be to the point and direct
- be lost or panic at unexpected inputs or alternatives	+ acknowledge input, indicate consequences (facts based)

Figure 9.11   How to cope with managerial dominance.

The impatience and action orientation of managers makes them very dominant, with the risk that they take over the meeting or presentation. Figure 9.11 shows a number of these risks and the possible countermeasures:

Managers hijacking the meeting can be prevented by maintaining the lead as presenter.

Build up tensions by withholding facts or solutions, but be to the point and direct. For example, it can be wise to start with a summary of the main facts and conclusions so that the audience knows where the presentation is heading.

Be lost or panic at unexpected inputs or alternatives. Most managers are fast and have a broad perspective that helps them to come forth with unforeseen options. Acknowledge inputs and indicate the consequences of alternatives as far as you can see them (fact-based!).

An example of an unexpected input might be to outsource a proposed development to a low-cost country. The outsourcing of developments of core components might require lots of communication and traveling, creating costs. Such a consequence has to be put on the table, but refrain from concluding that it is (im)possible.

EXERCISES

IN CLASSROOM FOR STUDENTS WITH WORKING EXPERIENCE

Bring a clear *architecture message* to a *Management team* at least two hierarchical levels higher than your supervisor, with *10 minutes* for *presentation, including discussion* (there is no limit to the number of slides).

An *architecture message* = *technology* options in relation to *market* and *product*. Address the *concerns* of the *management stakeholders*: translation is required from *technology* issues into *business consequences* (months, effort in person years, turnover, profit, investments).

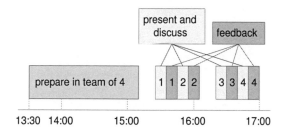

Figure 9.12 Example timeline of Management Presentation Exercise.

Take an actual and hot subject from today's practice. This is an ideal opportunity to conduct the presentation of an issue where you need management attention. Do pick an issue in the core of architecting, avoiding subjects that are clearly allocated to others, such as project management and marketing issues.

Figure 9.12 shows an example of the typical timeline for this in class exercise.

"HOME" WORK FOR STUDENTS WITH WORKING EXPERIENCE

Improve the presentation made in the classroom by adding facts and analysis. Present the improved version to your supervisors and ask them for feedback.

IN CLASSROOM FOR STUDENTS WITHOUT WORKING EXPERIENCE

Bring a clear *architecture message* to the *Board of Management*, with *10 minutes* for *presentation, including discussion* (there is no limitat to the number of slides).

An *architecture message* = *technology* options in relation with *market* and *product*. Address the *concerns* of the *management stakeholders*: translation is required from *technology* issues into *business consequences* (months, effort in person years, turnover, profit, investments).

10 Human Side

10.1 THE HUMAN SIDE OF ARCHITECTING

10.1.1 INTRODUCTION

Systems architecting involves much more than understanding technology and using technologies to create systems. Systems architects are working for, and are working with, humans. Architects are confronted continuously with human aspects. These human aspects might get lost in the hectic world of technology-oriented Product Creation. The technical origin of most of the design and implementation work lures designers into a technology-only viewpoint.

Human aspects cover a broad field that, in the academic world, is covered by the human sciences. Human sciences approach knowledge significantly differently from engineering sciences; it is a much "softer" world than the "hard" engineering world.

We will discuss the breadth of the human sciences and their relevance to systems architecting. The goal is to make (potential) systems architects aware of the importance of human aspects and to stimulate them into investing time in studying the human sciences.

We focus on the relevance of human aspects to systems architects, but most information and insights are applicable to engineers, designers, and managers.

10.1.2 HUMAN ASPECTS

Figure 10.1 shows an overview of human aspects as a two-dimensional space. One axis is the cultural diversity (vertical). The other axis is the amount of humans involved, starting with one individual and ranging to the entire society. The space of human aspects is covered by a range of human sciences, such as psychology and sociology, shown at the bottom.

Individual

Examples of attributes related to an individual are identity, self-perception, attitudes, physical condition, and health. *Psychology* focuses on the psyche of the individual, related to psychological aspects. *Psychiatry* copes with pathological psychological characteristics, such as personality and learning disorders. *Physiology* captures the knowledge of the physical aspects of humans. *Ergonomics* combines physical and mental human aspects. *Medicine* copes with pathological physical characteristics.

Traditionally, ergonomics is the main expertise area that is seen as relevant to systems architects. However, systems architects will meet all of these human aspects both in the company while cooperating with others and with external stakeholders.

For example, in security such straightforward biometrics can play a role. Biometrics might be disturbed by illnesses or physical handicaps. Security measures might

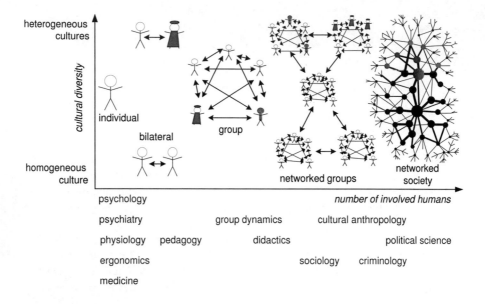

Figure 10.1   Overview of human aspects.

not work well because the measures do not fit with psychological needs. Security design, finally, has to consider mental disorders, too.

Bilateral

When two individuals meet, then they need to be able to understand each other, to communicate, and to behave such that both feel well and respected. Typical bilateral skills that are required whenever two individuals meet are active listening, empathy (the ability to feel or assess the emotions of the other), capability to express ideas, to give feedback, and to provide direction.

Most of these skills are fundamental in group interactions as well. Bilateral skills are the foundation of successful interaction in broader groups and networks.

An example of the value of bilateral skills is a situation where a designer has a conflict with the partner at home. The design discussion between architect and designer does not work well despite good ideas and suggestions from the architect. In this case, the architect has to discover that the current problem is not in the design and the discussion about the design. The actual problem is outside the immediate context: the conflict at home. By combining bilateral skills, the discussion might be postponed to a more suitable moment.

A specific subset of bilateral interaction is covered by *pedagogy*: how to educate children. Understanding of pedagogy can help in the understanding of bilateral relations.

Groups

Systems architects spend a significant amount of time in groups, for instance, in design and specification meetings, ad hoc task forces, strategy workshops, or reviews. "Group Dynamics" describe interaction between group participants.

Architects can function better in groups or teams when they understand the behavior of individuals. Many role models can help to understand roles that are required in teams and roles that fit specific individuals.

Networked Groups

When more and more individuals are involved, then there are many interpersonal relations. We can view that as networks or networked groups. *Sociology* studies how larger groups of humans live, behave, and cooperate. *Didactics* focuses on teaching to larger groups.

In the example of security, we can also see the need to understand social aspects. Many security problems originate from social behavior. For instance, malware makers apply social engineering to penetrate secure systems. Social engineering uses expected social behavior to harvest confidential information.

Networked Society

Today's society contains globally about 10 billion individuals. Global society can be viewed as a huge network. In larger populations, humans start to show political behavior: using power and coalition strategies to achieve personal or other local goals.

Most systems architects dislike politics intensely. Politics operates contrary to the natural architecting style: trying to find a solution that maximally satisfies stakeholders, based on facts and figures. The systems architect is the catalyst to be fact- and task-driven in groups, and to discuss the content, rationales, and solutions instead of compromising and polluting the whole by narrow personal interests.[1]

Another phenomenon that pops up in larger populations is crime committed by people who have chosen to operate outside the social system and ignoring legal rules.

Heterogeneous Cultures

The vertical axis shows cultural diversity. Culture consists of unwritten rules that very slowly emerge in a population. These rules are ingrained in all individuals of the population. In due time, the rationale of the rules is lost, but the population continues to live according to these rules. Changing culture is a tedious and slow process.

Cultural anthropology studies the cultural aspects of populations. The cultural background of individuals plays from individuals to the entire society. The cultural

---

[1]Note that personal interests need to be acknowledged and taken into account, as described in the bilateral skills subsection. However, acknowledging and taking into account is not the same as fulfilling.

background of an individual shapes the beliefs and behavior of an individual. In-teraction between individuals with different backgrounds may have unexpected side effects.

For example, Dutch people are quite blunt and not hierarchical oriented. In Dutch culture, an employee may contradict the boss. When a Dutch employee contradicts an American manager higher up in the hierarchy, then the American manager may be offended by the contradiction.

Cultural differences are not limited to geographical boundaries or ethnological backgrounds. Companies (e.g., IBM, Google, Apple, or Microsoft) do have specific cultures. Disciplines (e.g., software engineering, electronics engineering, or sales) can have specific cultures. Any group of people gradually develops an own culture.

10.1.3 HUMAN CONTEXT

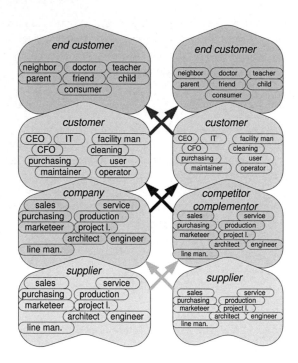

Figure 10.2 The systems architecting context, shown here as value chains, is full of human stakeholders.

Systems architecting is taking place in a context full of human players. Figure 10.2 shows the value chain where most of the systems architecting takes place in the *company* part. In the company itself, there are many human stakeholders. In addition, suppliers, customers, and end-customers consist of many human stakeholders.

Note that, in most processes,[2] an abstraction of the stakeholder is used, such as *customer, consumer, user, employee*, etc. The needs of these abstracted stakeholders are captured in other abstractions such as requirements and specifications. Architects need to be aware of the rich variations in humans hidden behind these abstractions.

For instance, a specification might indicate that a product is targeted at elderly citizens. "Elderly citizens" is much more abstract than "85-year-old mister Smith who cannot find his remote control that is so small that it always disappears."

Systems architects interact with external and internal stakeholders. Quite often it is impossible to know all of them personally, forcing architects to work more indirectly and to apply abstractions. For instance, Sales and Marketing Managers meet much more customers and often represent them during requirement capturing. Systems architects should at least meet a few "live" customers. They need to balance the degree of abstraction and the amount of attention to internal and external stakeholders.

10.2 FUNCTION PROFILES: THE SHEEP WITH SEVEN LEGS

10.2.1 INTRODUCTION

Many human resource and line managers struggle with the following questions:

- What people have the potential to become good systems architects?
- How to select (potential) systems architects?

Employees thinking about their careers might similarly wonder if they have the capabilities to become a good systems architect.

We list a number of characteristics of individual humans. We map these characteristics on different jobs, such as systems architect, developer, and line manager, indicating the relative importance of this characteristic to that job. We first discuss the different jobs and their typical characteristics in Subsections 10.2.2 to 10.2.7. Then we elaborate the characteristics in Subsection 10.2.8.

10.2.2 SYSTEMS ARCHITECT PROFILE

The profile of the "ideal" system architect shows a broad spectrum of required skills. Quite a bit of emphasis in the skill set is on *interpersonal skills, knowledge*, and *reasoning power* (Figure 10.3).

This profile is strongly based on an architecting style of technical leadership, where the architect provides direction (*knowledge* and *reasoning power*) as well as moderates the integration (*interpersonal skills*).

The required profile is so demanding that not many people fit into it; it is a so-called **sheep with seven legs**. In real life, we are quite happy if we have people

---

[2]A perfect example is this section itself, where we used several abstractions to discuss humans.

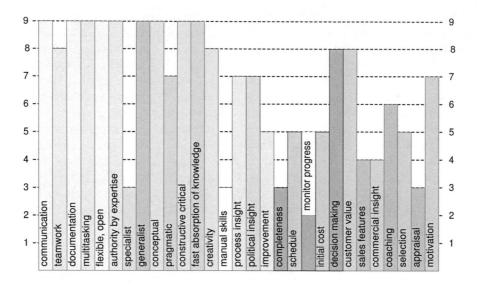

Figure 10.3　The function profile of the systems architect.

available with a reasonable approximation to this profile. The combination of complementary approximations of such an ideal architect allows for the formation of architecture teams. Such a team of architects can come close to this profile.

Most Discriminating Characteristics

In practice, the following characteristics are quite discriminating when selecting (potential) systems architects:

Generalist The first reduction step is to select the *generalists only*, reducing the input stream with one order of magnitude. The majority of people feel more comfortable in the specialist role.

Multitasking The next step is to detect those people that need undisturbed time and concentration to make progress. These people become unnerved in the job of the systems architect, where frequent interrupts (meetings, telephone calls, or people walking in) occur all the time. Ignoring these interrupts is not recommendable; this would block the progress of many other people. Whenever people with poor multitasking capabilities become systems architects, they are in severe danger of stress and burnout. Hence, it is also beneficial to the person's self to assess the multitasking characteristic fairly.

Authority by expertise The attitude of the (potential) architect is important for long-term effectiveness. Architects who work based on delegated *power* instead of *authority by expertise* are often successful in the short term, creating a single focus in the beginning. However, in the long run, the inbreeding of ideas takes its toll. Architecting based on knowledge and contribution (e.g., *authority by exper-tise*) costs a lot of energy, but it pays back in the long term.

Conceptual thinking and pragmatism The balance between conceptual thinking and being pragmatic is also rather discriminating. Conceptual thinking is neces-sary for an architect. However, the capability to translate these concepts into real world activities or implementation is crucial. This requires a pragmatic approach. Conceptual-only people dream up academic solutions.

Constructive critical or critical thinking capability. A good systems architect is nearly obsessive in the need to understand problems and solutions. Many engineers are too quickly satisfied with an answer. However, critical questioning should not transform into complaining; systems architects need to be solution and improve-ment oriented.

10.2.3 TEST ENGINEER PROFILE

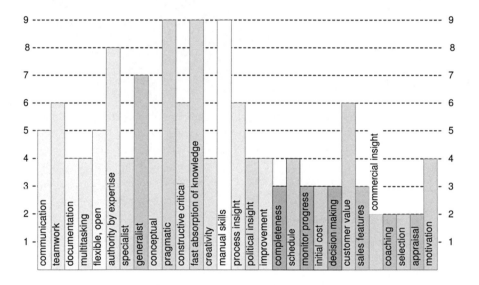

Figure 10.4 The function profile of the test engineer.

The *test engineer* function at the systems level requires someone who *feels* and *understands* the system. Test engineers are capable of operating the system fluently and know its quirks inside out (Figure 10.4).

The main difference between an architect and a test engineer is the difference in balance between **conceptual thinking** and **practical doing**. Test engineers often have an excellent intuitive understanding of the system; however, they tend to lack conceptual expression power and the communication skills to use this understanding proactively. For instance, test engineers find it difficult to lead the design team.

10.2.4 DEVELOPER PROFILE

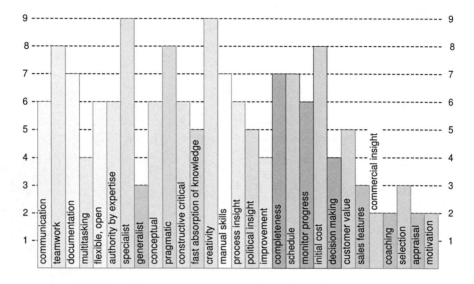

Figure 10.5 The function profile of the developer.

The core value of developers is their specific discipline knowledge. Good developers excel in a limited set of specialties, knowing all tricks of the trade. On top of this, they should be able to deploy this knowledge in a creative way. In today's large development teams, a reasonable amount of *interpersonal skills* are required as well as *reasoning power* and *project management* skills (Figure 10.5).

10.2.5 OPERATIONAL LEADER PROFILE

The *operational leader*, for instance, a project leader or program manager, is totally focused on the result. This requires *project management* skills, the core discipline for operational leaders.

The *multitasking* capability is an important prerequisite for the operational leader too. If this capability is missing, the person runs a severe risk of becoming a case of burnout.

Note also that the operational leader functions as a kind of gatekeeper, where the *completeness* is important (Figure 10.6).

10.2.6 LINE MANAGER PROFILE

The *line manager* manages the intangible assets of an organization: people, technology, and processes. Technology and process knowledge are tightly coupled with people; this knowledge largely resides in people and is deployed by people. *Human resource management* skills and *process* skills are the core discipline for line

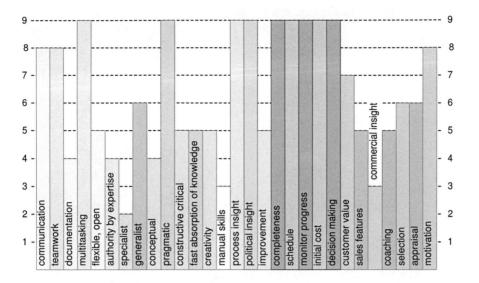

Figure 10.6   The function profile of the operational leader.

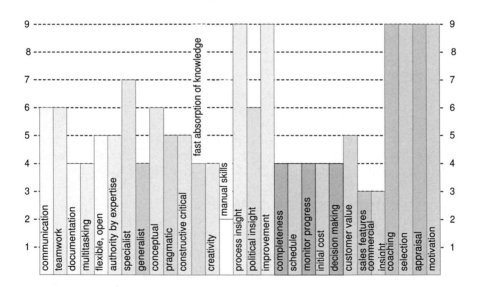

Figure 10.7   The function profile of the line manager.

managers, which need to be supported with sufficient *specialist* knowledge. Line managers are typically owners of the People, Process, and Technology Management Process (Figure 10.7).

10.2.7   COMMERCIAL MANAGER PROFILE

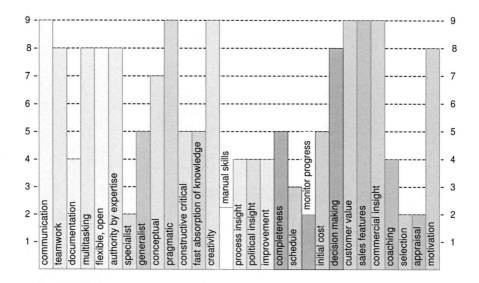

Figure 10.8   The function profile of the commercial manager.

The *commercial manager* needs a commercial way of observing and thinking. This way of thinking appears to be fuzzy and not logical for technology-oriented people. From a technology-oriented perspective, a strange *mind warp* is required to perform a commercial manager function.

The commercial manager is a valuable complement to the other functions, responsible for aspects such as salability and value proposition. Typical commercial managers are marketing managers and product managers (Figure 10.8).

10.2.8   DEFINITION OF CHARACTERISTICS

Interpersonal skills

Communication The ability to communicate effectively. Communication is a two-way activity: presenting information as well as receiving information is important.

Teamwork The ability to work as member of a team in such a way that the team is more than the collection of individuals.

Documentation The ability to create clear, accessible and maintainable documentation in a reasonable amount of time.

Multitasking The ability to work on many subjects concurrently, where (frequent) external events determine the task-switching moments.

Flexible, open The attitude of respecting the contributions of others, the willingness to show all personal considerations, even if these are very uncertain, the willingness to adopt the solutions of others even in case of strong personal opinions.

Note that this overall attitude does not mean that a flexible and open person always adopts the ideas of others (chameleon behavior). The true strength of this characteristic is to apply it when necessary, so adopt an alternative solution if it is better.

Authority by expertise The personality that convinces people by providing data, instead of citing formal responsibilities. Hard work is required before authority by expertise is obtained; a good record of accomplishment and trust has to be built up. Authority is earned rather than enforced.

Knowledge

In terms of characteristics, knowledge is qualified in two categories: generalists and specialists.

Generalists Persons that are always interested in the neighboring areas: How does it fit in the context? How does the "whole" work?

Specialists Persons that are always interested in knowing more detail.

Reasoning Power

Conceptual The ability to create the overview, to abstract the concepts from detailed data. The ability to reason in terms of concepts.

Pragmatic The ability to accept nonideal solutions, to go after the 80% solution. The ability to connect "fuzzy" concepts to real-world implementations.

Constructive critical The ability to identify problems, to formulate the problems, and to trigger solutions. The term *critical thinking* is also used. Note that critics serve a constructive goal: to achieve better results.

Fast absorption of knowledge The ability to jump into a new discipline and to absorb the required knowledge in a short time. Systems architects are never able to know all about the technologies used in the systems. This capability helps them to get the right knowledge when needed.

Creativity The ability to come up with new, original ideas. A specific subclass of this ability is lateral thinking: applying knowledge from entirely different areas on the problem at hand.

Executing Skills

Manual Skills The ability to **do** things, for instance, building or testing something. This ability is complementary to the many "mental" skills in this list of characteristics.

Process Skills

Process insight The ability to understand specific processes, the ability to recognize de facto processes, the ability to asses formal and de facto processes, both the strong points as well as the weak points.

Political insight The ability to recognize the political factors: persons, organizations, motivations, power. The ability to use this information as neutralizing force or "depoliticizing": facts- and objectives-based decision making instead of power-based decision making.

Improvement drive The ever-present drive to improve the current situation, never getting complacent.

Project Management Skills

Completeness The ability to pursue **all** information. This is often done by means of spreadsheets or databases. Large collections of issues are maintained and processed.

This ability is often complementary to, or even conflicting with, the ability to create understanding and overview: the parts view versus the holistic view.

Schedule The ability to create schedules: activities and resources with their relationships, scheduled in time.

Monitor progress The ability to monitor progress, the ability to chase people, and the ability to find and resolve the causes of delays.

Initial cost The ability to create initial cost estimates and to refine these into budgets. The ability to understand and reason in terms of initial costs. Initial costs are the one-time investments needed to develop new products and or businesses.

Decision making The ability to make choices and to handle the consequences of these choices.

Commercial Skills

Customer value The ability to see and understand the value of a product or service for a customer. The ability to asses the value for the customer.

Sales feature The ability to recognize features needed to sell the product. The ability to characterize the relevant characteristics of these features ("tick-mark only," "competitive edge," "show-off," etc.).

Commercial insight The ability to think in commercial terms and concepts, ranging from "branding" to "business models."

Human Resource Management Skills

Coaching The ability to coach other people; help other people by reflection, by stimulating independent thinking and acting.

Selection The ability to select individuals for specific jobs. The ability to interview people and to assess them.

Appraisal The ability to assess employees and to communicate this assessment in a fair and balanced way.

Motivation The ability to make people enthusiastic, to motivate them beyond normal performance.

10.3 INTERPERSONAL SKILLS

10.3.1 INTRODUCTION

We often take for granted that two individuals can cooperate. However, in practice, many problems arise at the fundamental level of cooperation between two individuals. We will discuss the wonder of communication in this section since communication is the starting point for cooperation.

We encourage architects to develop their interpersonal skills further. There are many courses where methods and techniques are offered that can be deployed between two (and often more) individuals, for example[3],

- Investigation by questioning and paraphrasing
- Constructive and balanced feedback
- Conflict management
- Appraisal and recognition of the contribution of others
- Facilitating creativity

10.3.2 THE WONDER OF COMMUNICATION

If someone wants to transfer an idea to another person, then this idea is encoded in a message. This message is encoded by a variety of means, ranging from the verbal message to the nonverbal message such as facial expressions, gestures, and voice modulation. The encoding of this message depends on many personal aspects of the *speaker*; see Figure 10.9. The receiver of this message has to decode this message and interpret the message, however, based on many similar personal aspects of the *receiver*.

From the technical point of view, a pure miracle is happening in communication: sender and receiver use entirely different configured encoders and decoders and nevertheless we humans, are able to convey messages to others.

The mechanism behind this miracle can be understood by extending the model of sender and receiver as in Figure 10.10. The mutual understanding is built up in an interactive calibration process. By phrasing and rephrasing examples, illustrations and explanations, the coding and decoding information is calibrated. The capability to communicate this way is called *active listening*.

---

[3]This is a free interpretation of the interpersonal skills taught by Hay Management Consultants in 1997.

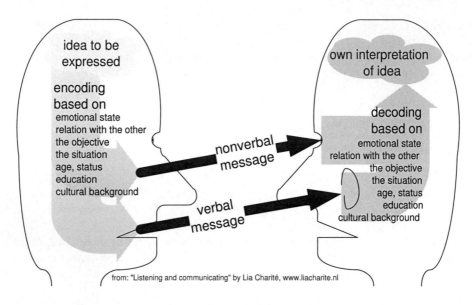

from: "Listening and communicating" by Lia Charité, www.liacharite.nl

Figure 10.9   Active listening: the art of the receiver to decode the message.

to calibrate:
repeat many times with different
examples, illustrations, and explanations

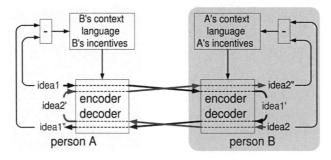

Figure 10.10   Intense interaction needed for mutual understanding.

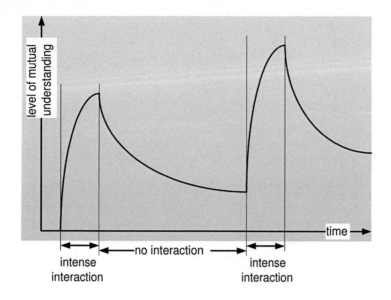

Figure 10.11   Mutual understanding as function of time.

Calibration information is very dynamic; part of the coding depends on volatile issues, such as mood and context. During interaction, mutual understanding improves since continuous calibration takes place. Without interaction, mutual understanding degrades due to the dynamics of interpretation. Figure 10.11 visualizes mutual understanding as function of time and interaction.

Glossaries of terms, unified notations, and all these kind of measures do not fundamentally address the communication difficulties explained here. In fact, standardized terminology and notations are a minor factor[4] in comparison with the human differences that have to be bridged continuously.

10.4   TEAMWORK

10.4.1   WHY WORK IN TEAMS?

Today's product creation projects involve so many fields of expertise that we need many different specialists. Teams are a way to organize the project. From management perspective, manageability is one of the main reasons for using teams.

Teams can also be an effective way to benefit more from the strengths of individuals within a group. A well-designed team is more than its constituting members. Figure 10.12 shows a schematic version of the figure of the three apes, where one

---

[4]Dogmatic applied unification of terms and notations often works in counterproductive ways. Problems or viewpoints might be more easily expressed in other terms, while the unification drive blocks the search for a mutually understandable expression. Active participation is required to obtain understanding.

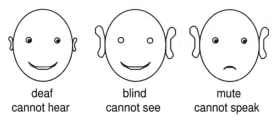

deaf   blind   mute
cannot hear   cannot see   cannot speak

but in the team two can hear,
two can see, and two can speak

Figure 10.12   Teams consist of peoples with complementary skills and knowledge.

ape cannot see, one ape cannot hear, and one ape cannot talk. If all three apes team up, then we have a team where two apes can see, two can hear, and two can talk. The team members are complementary, and jointly they can do much more than each individual alone.

10.4.2   TEAM SIZE

Teams can vary from very small (two people) to very large (thousands of people). However, large teams are often further divided into smaller teams since human interaction becomes more difficult with more team members. For example, an entire project is decomposed into subsystems and further decomposed into components, where the team organization follows the same decomposition.

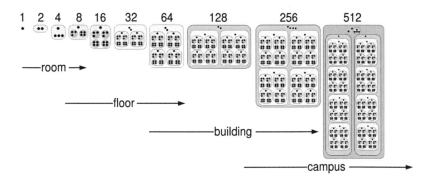

Figure 10.13   The way of working of a team depends on the team size. Every factor 2 in size creates a different paradigm.

In practice, only small teams, for example, between three and eight people, show intense interaction. When teams are bigger, the team de facto divides itself further into "human" scale teams. Figure 10.13 shows that large organizations are broken down into smaller units. It also shows that size affects the housing and location of teams.

The operation of a team depends strongly on its size. Very small teams of two or three people will interact more intensely and informally than larger teams. Larger teams will need more time to communicate, lowering their efficiency. In fact, every increase of the organization's size of circa 50% triggers changes in the mode of operation.

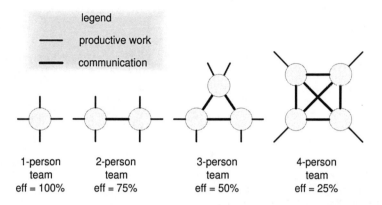

Figure 10.14   A very simplistic model of team communication and productivity.

Figure 10.14 shows a very simplistic model of the communication and productivity of a team member. Every team member is modeled as being able to spend time on four tasks each 25% of the time. Every task is either producing something or communicating with someone. This simple model shows that working with more people gets quickly less efficient due to communication overhead.

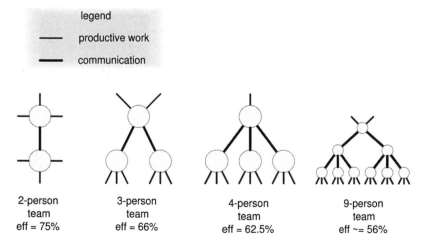

Figure 10.15   A very simplistic model of team communication and productivity with hierarchy.

This simple model can be transformed into a hierarchical team model; see Figure 10.15. The hierarchy reduces the communication overhead, and hence, improves efficiency. Note that the hierarchy increases the length of the communication paths. Longer communication paths suffer from more latency and more deformation. A network organization combines a hierarchical structure and more direct communications between individuals, creating an effective organization.

10.4.3 TEAM COMPOSITION

A good team consists of complementary members, where the members can cooperate well. The team "chemistry" must be good. In the literature, many role models are given that can be used to "design" a team.

plant creative	*team worker* cooperative, averts friction	*implementer* disciplined, conservative, doer
resource investigator enthusiastic communicator	*shaper* driver, dynamic	*completer finisher* conscientious, painstaking
coordinator mature, chairman	*monitor evaluator* sober, analytical	*specialist* single-minded, rare skills

Belbin's team roles

Figure 10.16 Belbin team roles.

Figure 10.16 shows the roles described by Meredith Belbin [3]; see [1] for a summary of these roles. People tend to have preferred role patterns that they follow. This is not black and white; for instance, people can be mostly *plant* and somewhat of a *chairman* at the same time. The idea is that a good team needs all different roles.

Six thinking hats by Edward de Bono

Figure 10.17 Six thinking hats by Edward de Bono.

Edward de Bono [5] provides another role model, the so-called Six Thinking Hats. These colored hats symbolize the natural attitude that persons bring into the team. Figure 10.17 shows the six different colors and their main characteristics. Again, the idea is that these different kinds of people are complementary—a team needs positive-and negative-oriented members, creative and process-oriented members, and neutral and feeling-oriented members. Note that some members can take on all colors: they are chameleons.

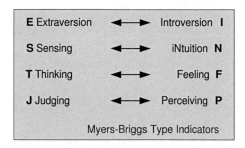

Figure 10.18 Myers–Briggs type indicators.

The Myers–Briggs type indicators [15] provide a famous ontology. Four characteristics with their two extremes are used as parameters to classify people, following Jung's ideas with similar characteristics. In Figure 10.18, these four characteristics are shown: *Extraversion* and *Introversion*, *Sensing* and *iNtuition*, *Thinking* and *Feeling*, and *Judging* and *Perceiving*. Someone's personality type can be captured by concatenating the four letters of these characteristics. For example, it is often observed that systems architects have INTP as personality; see, for instance, http://www.e-mbti.com/intp.php.

To design a team the roles/personalities of its intended participants have to be taken into account. However, we also have to design the team to get

- A multitude of opinions
- Coverage of the involved stakeholders
- Coverage of knowledge and skills

10.4.4 THE PROCESS OF CREATING AND EMPLOYING A TEAM

Let us assume that someone in the organization has the need for a team, for example, the business manager or project leader. We will call this person the team owner, meaning that this person feels responsible for a good functioning team and has a need for its results. Figure 10.19 shows this team owner at the top of the figure.

This team owner will compose the team as described in Subsection 10.4.3, will arrange facilities, such as housing as described in Subsection 10.4.5, and will have to provide a charter for the team. A charter provides the scope for the team: *what* has to be done *why*, *how* to achieve this, *whom to involve*, and *when* and *where*. This charter gives direction to the team. However, the team should be able to determine its own way-of-working within the charter. Micromanagement of the team by its owner or

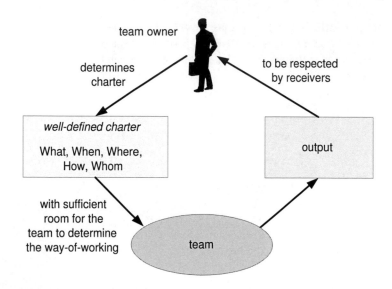

Figure 10.19 The process of creating and employing a team.

others outside the team will greatly reduce the team's productivity. In other words, the team has to be empowered by the owner.

The team will produce results while working. These results have to be respected by the receivers, such as the owner. If the results are not respected, then this will discourage this team and the successive teams: *Why engage in team activity if the results are not taken seriously?* The owner should consider this aspect before initiating a team. Teams might reach conclusions that are undesired by the initiators. Sometimes, teams are initiated as a decoy rather than a real goal; this works once; after that, employees become frustrated, and it takes a long time to repair trust and motivation.

10.4.5 HOUSING AND LOCATION

Housing can be used as an instrument to boost team productivity and cohesion. When team members sit in the same room, then they will communicate more frequently and more naturally.

Figure 10.20 shows an example of a room used by the systems engineering team that designed the next-generation wafer steppers at ASML. This large room had space for the normal desks with PCs, and a meeting space with plenty of wall space. The wall space can be used for white-boards, flipcharts, and large format printouts of diagrams. This type of war room is very effective.

Housing is an effective means to improve team efficiency. Team members should be able to communicate without any obstacles, such as distance. The room described above also supports communication by providing wall space and simple means to share visualizations. In LEAN product development also, the physical system or the

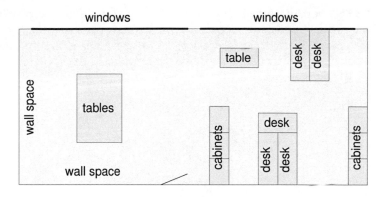

Figure 10.20   "War Room" is very effective.

components are shared in the same location, supporting tactile and visual discussions. In a Concurrent Design Facility, the room facilitates sharing of computer-based models by using multiple large screens for projection.

10.4.6   CONCURRENCY

Organizations seem to increase the number of activities continuously. Individual employees get more and more activities they have to perform, often more or less concurrently. These activities tend to fragment the time of individuals, working a little on the first activity, do something on the second activity, continue with the next, etc. Figure 10.21 shows this fragmentation. It also shows that working in burst-mode, for example, working focused for one day, one week, or one month on a single activity, can be more efficient because less context switches are required.

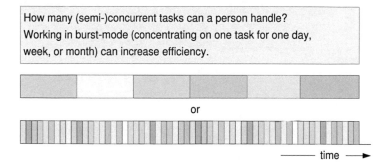

Figure 10.21   Many engineers have to divide their attention over a multitude of activities.

Figure 10.22 takes this line of reasoning further. It shows six activities that are being executed in parallel and the same activities, but now executed sequentially. When the activities are done in parallel, then the results of all activities become available when all work is finished. When the activities are performed sequentially, then the first result gets available after one sixth of the time of the parallel approach. This means that the result itself, and feedback on this results becomes available much earlier. In other words, when working in parallel all results are late.

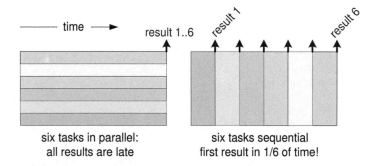

six tasks in parallel:
all results are late

six tasks sequential
first result in 1/6 of time!

Figure 10.22   It can be more efficient to perform activities sequentially rather than in parallel.

Figure 10.22 is a simplification of reality. It might be that some activities need to be parallel, for example, because of the inherent elapsed time of the activities. Nine mothers cannot deliver one baby in one month, although nine mothers can deliver nine babies in nine months.

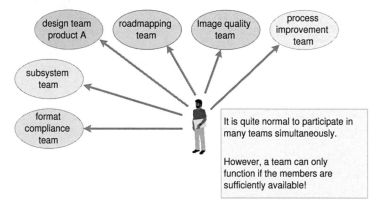

Figure 10.23   One person will be a member of multiple teams.

If we map these considerations on teams, then we should realize that, in practice, people, and certainly systems architects, participate in multiple teams at the same time. An example is shown in Figure 10.23.

The systems architect in Figure 10.23 works on the creation of a product and is part of the *design team* of that product. At the same time, the architect participates in more specific product creation teams, such as one of the *subsystems* and some specific aspects such as *format compliance*. However, systems architects will also participate in broader concerns of the organization, such as *image quality*, *roadmapping*, and *process improvement*.

If all these activities run concurrently, then some of them might suffer. For example, it might be better to focus for 4 weeks entirely on roadmapping together with the other roadmapping team members and, after that time, continue with the day-to-day architecting concerns of product creation.

10.4.7   CRITICAL SUCCESS FACTORS

Figure 10.24 shows a summary of the critical success factors that we have discussed.

well defined charter

clear owner of the result

respect for the output of the team

freedom of way-of-working

housing and location

availability of team members

complementary roles

diversity, pluriformity

Figure 10.24   critical success factors for teams.

Well-defined charter   for the team: *what*, *where*, *when*, *how*, *whom*.

Clear owner of the result   Who needs the team, who ensures appropriate support and facilities, who is worried when the team hits obstacles?

Respect for the output of the team   The team with all its expertise might draw unwanted, undesired conclusions. The output of the team has to be respected; otherwise, follow-up teams will not be motivated.

Freedom of way-of-working   The owner has to empower the team; micro-management will stifle the team's effectiveness.

Housing and location   are instruments to forge a team; co-location in a war room is recommended.

Availability of team members   is a prerequisite to work as a team. Availability can be arranged by careful preparation and timely allocating team members.

Complementary roles   so that the team is more than the sum of the individual participants.

Diversity, multitude of opinions   Some controversy or tension in the team is healthy to prevent inbreeding and blind spots.

EXERCISES

IN CLASSROOM FOR STUDENTS WITH WORKING EXPERIENCE

Make a lightweight self-assessment using the characteristics of Section 10.2; rank yourself on every characteristic on a scale from 1 to 9.

Reflect on the following questions:

- What characteristics do you find difficult to rank?
- Are there characteristics where your own assessment is likely to be quite different from the assessment of others?
- What characteristics would you like to improve?

Make also a lightweight assessment of one of your managers, using the same characteristics and scale.

IN CLASSROOM FOR STUDENTS WITHOUT WORKING EXPERIENCE

Write an one-page essay on the desired human interaction model in your company. Elaborate briefly on the following aspects:

- Relation between function and profile
- The culture, values, and attitudes that you would like to stimulate
- The rationale behind these choices

11 Reflection and Wrap-Up

11.1 REFLECTION APPLIED ON SYSTEMS ARCHITECTING

11.1.1 LEARNING AND REFLECTION

Systems architecting is a competence that people mostly develop in practice based on experience. This book has brought you a lot of systems architecting knowledge and insights. Potential systems architects need to link the knowledge to their experience to develop systems architecting competence.

Reflection facilitates learning by relating knowledge and experience; see Schön [19].

Reflection is an essential step of learning. Experiences are simply accumulating if no reflection is done. Reflection transforms experiences into insights and helps to develop capabilities. Reflection can be seen as an intelligent feedback mechanism applied on individuals.

Schön differentiates *Reflection In Action* (RIA) and *Reflection On Action* (ROA). RIA is reflecting concurrently with the action itself, while ROA is retrospective when the action has been finished. Note that *Reflection Before Action* can also be quite useful: thinking about the approach and the expected reactions is valuable as preparation but also sharpens the Reflections In and On Action.

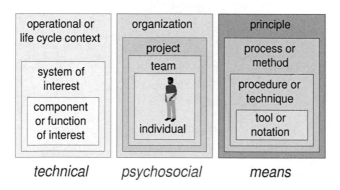

Figure 11.1 Examples of different scopes for reflection

Figure 11.1 shows several examples of the scope of reflection in the dimensions *technical*, *psychosocial*, and *means*. The scope can be very specific (component or function, the individual, or a tool or notation). The scope can be increased, for example, to look at the entire system-of-interest or to also include the operational and life-cycle context. Reflection in a narrow scope is more specific and more manageable. However, the impact of the reflection tends to be larger for larger scopes. We recommend starting small and specific and gradually increasing the scope of reflection.

11.1.2 HOW TO REFLECT

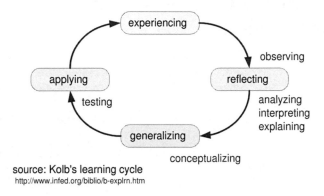

source: Kolb's learning cycle
http://www.infed.org/biblio/b-explrn.htm

Figure 11.2   Reflection Cycle.

We use the learning cycle as described by Kolb [10] to explain how to reflect; see Figure 11.2. Kolb's learning cycle is a simple model with four steps:

Experiencing specific situations in practice.
Reflecting on the experience. We can describe the situations based on observations. Next, we can analyze what has happened, interpreting the observations. After analysis, we can explain this specific situation.
Generalizing and conceptualization of this specific situation and previous experiences to achieve applicable insights for future use.
Applying the insights to test them in practice.

Someone who reflects steps out of a situation and tries to understand the situation by asking questions:

- What stakeholders are involved?
- What are their needs and concerns?
- What is our goal?
- How did we get in the current situation?
- What is going well, what is going bad?
- What approach can we take?
- What do we expect to happen?
- Others

11.1.3 REFLECTION REPORT

We encourage following the learning-cycle logic when writing a reflection report. It could contain the following:

Subject or goal of the reflection report.
Description of your experiences. Try to avoid interpretations in the description, limit yourself to observations (what did you see, hear, feel, etc.).
Analysis of your experiences; can you understand and explain what happened?
Lessons Learned and insights obtained from the reflection.
Actions as follow-up: what are you going to do with your new insights?

Avoid broad generic statements in the report (e.g., "Everybody was complaining"); try instead to illustrate with specific examples.

11.2 WRAP-UP

11.2.1 RED THREADS

We described the following recurring principles or fundamental concepts to systems architecting in the beginning of the book:

Communication between many stakeholders with different backgrounds
Understanding of problems and solutions to make appropriate choices
Providing overview and insight to all stakeholders
Awareness of unknowns and uncertainties in all aspects of systems architecture
Goals and means articulation
Customer and life-cycle context
Being sharp and factual rather than acting on beliefs
Feedback and iteration as leading principles in doing architecture work

These principles and concepts have been discussed implicitly or explicitly in the viewpoints on architecting:

- Process and Organization
- Role and Task of the Systems Architect
- From Customer Understanding to Requirements
- Systems Architect Methods and Means
- Strategy
- Harvesting Synergy, Product Families
- Supporting Processes
- Systems and Software
- Boardroom Presentation
- Human Side

A major challenge for students and readers is to apply insights from this book in their own working environment. We recommend introducing architecting methods and techniques gradually. Working by example, such as by showing the value of techniques and methods in practice, is often most effective. Apply these methods and techniques when there is a clear need. Working from the needs (demand driven) creates less opposition than pushing beliefs.

The process of introducting architecting, if done gradually, takes up much time. Hence, students and participants are urged to reflect often on theory and needs in practice. The red threads and the viewpoints can be used as checklists for reflection.

EXERCISES

IN CLASSROOM FOR STUDENTS WITH WORKING EXPERIENCE

Individual work: Make a short-term and long-term plan, both as single-page documents.

The short-term plan focuses on practical steps: what can you do in your company to improve systems architecting? For example, "For this project, I will make a key driver graph next month."

The long-term plan follows the ideas of roadmapping applied on your personal development in the context of your company. Personal development steps should be related to the needs of your company and trends in the domain.

Teamwork: Make a plan for the "home" work.

"HOME" WORK FOR STUDENTS WITH WORKING EXPERIENCE

Team work: Integrate all exercises in one presentation, showing the current state of systems architecting from all chapter viewpoints. Show and discuss this presentation with your company supervisor. Transform the presentation into a report. The report format allows you to provide a written explanation. The report should contain less than 20 pages.

Individual work: Finish the short- and long-term plan that you started to make in the classroom.

Write a reflection paper, one to two pages, about the course in relation to your working experience.

IN CLASSROOM FOR STUDENTS WITHOUT WORKING EXPERIENCE

Individual work: Iterate once more over all material and exercises. Make a personal list of highlights and main lessons learned.

Teamwork: Prepare an outline for a business report to be made as homework.

"HOME" WORK FOR STUDENTS WITHOUT WORKING EXPERIENCE

Teamwork: Integrate all answers into one business report. The business report should help convince outside investors that you know your business, customers, processes, technology, and people, and that it is therefore a good investment.

Individual work: Write a one- to two-page reflection report about the course and homework.

References

Belbin Associates. Belbin team-role summary sheet. `http://www.belbin.com/content/page/49/Belbin_Team_Role_Descriptions.pdf`, 2001.

Kent Beck. *Extreme Programming Explained: Embrace Change*. Addison-Wesley, Reading, MA, 2000.

Meredith Belbin. *Management Teams, Why They Succeed or Fail*. Butterworth-Heinemann, Boston, MA, 1981.

Daniel Borches. *A3 Architecture Ovevriews: A Tool for Effective Communication in Product Evolution*. Ph.D. thesis. Wohrmann Print Service, Enschede, Netherlands, 2010.

Edward de Bono. Six thinking hats. `http://www.debonogroup.com/six_thinking_hats.php`.

Jean-Marc DeBaud and Klaus Schmid. A systematic approach to derive the scope of software product lines. In *21st international Conference on Software Engineering: Preparing for the Software Century*, pages 34–47. ICSE, 1999.

Thomas Gilb. *Competitive Engineering: A Handbook For Systems Engineering, Requirements Engineering, and Software Engineering Using Planguage*. Elsevier Butterworth-Heinemann, London, 2005.

Ivar Jacobson, Grady Booch, and James Rumbaugh. *The Unified Software Development Process*. Addison-Wesley, Reading, MA, 1999.

Ivar Jacobson, Martin Griss, and Patrik Jonsson. *Software Reuse; Architecture, Process and Organization for Business Success*. ACM Press, New York, 1997.

D. A. Kolb. *Experiential Learning*. Prentice Hall, Upper Saddle River, NJ, 1984.

Klaus Kronlöf, editor. *Method Integration; Concepts and Case Studies*. John Wiley, Chichester, England, 1993. A useful introduction is given in Chapter 1, The Concept of Method Integration.

James Morgan. Applying lean principles to product development. `http://www.sae.org/manufacturing/lean/column/leanfeb02.htm`, 2010.

Gerrit Muller. Architectural reasoning explained. `http://www.gaudisite.nl/ArchitecturalReasoningBook.pdf`, 2002.

Gerrit Muller. CAFCR: A multi-view method for embedded systems architecting: Balancing genericity and specificity. `http://www.gaudisite.nl/ThesisBook.pdf`, 2004.

Isabel Myers. *The Myers-Briggs Type Indicator*. Consulting Psychologists Press, Palo Alto, CA, 1962.

Henk Obbink, Jürgen Müller, Pierre America, and Rob van Ommering. COPA: A component-oriented platform architecting method for families of software-intensive electronic products. `http://www.hitech-projects.com/SAE/COPA/COPA_Tutorial.pdf`, 2000.

William H. Press, William T. Vetterling, Saul A. Teulosky, and Brian P. Flannery. *Numerical Recipes in C: The Art of Scientific Computing*. Cambridge University Press, Cambridge, England, 1992. Simulated annealing methods page 444 and further.

Eberhardt Rechtin and Mark W. Maier. *The Art of Systems Architecting*. CRC Press, Boca Raton, Florida, 1997.

D. Schön. *The Reflective Practitioner. How Professionals Think In Action*. Basic Books, New

York, 1983.

Carnegie Mellon Software Engineering Institute SEI. Software engineering management practices. `http://www.sei.cmu.edu/managing/managing.html`, 2000.

Wikipedia. Rational unified process (rup). `http://en.wikipedia.org/wiki/Rational_Unified_Process`, 2006.

Wikipedia. Rup test discipline. `http://en.wikipedia.org/wiki/Rational_Unified_Process#Test_Discipline`, 2006.

Pictorial Index

The electronic version of a picture can be found by prefixing the picture name with "http://www.gaudisite.nl/figures/" and post-fixing it with ".html" for picture information or with ".wmf" for the picture in Windows Media Format (wmf). Pictures in wmf format can be inserted and scaled in Microsoft Office programs such as PowerPoint and Word. Use of all images is allowed when the source is referred to.

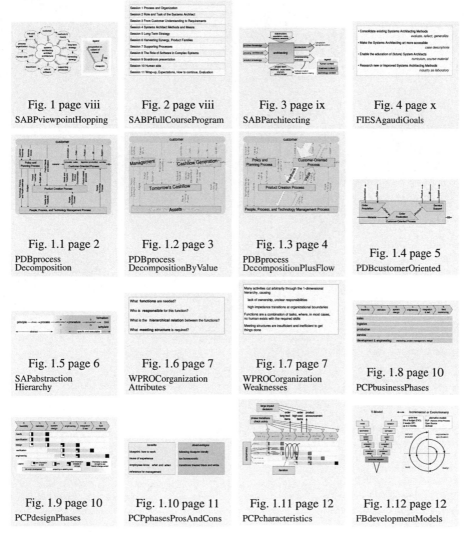

Fig. 1 page viii
SABPviewpointHopping

Fig. 2 page viii
SABPfullCourseProgram

Fig. 3 page ix
SABParchitecting

Fig. 4 page x
FIESAgaudiGoals

Fig. 1.1 page 2
PDBprocess
Decomposition

Fig. 1.2 page 3
PDBprocess
DecompositionByValue

Fig. 1.3 page 4
PDBprocess
DecompositionPlusFlow

Fig. 1.4 page 5
PDBcustomerOriented

Fig. 1.5 page 6
SAPabstraction
Hierarchy

Fig. 1.6 page 7
WPROCorganization
Attributes

Fig. 1.7 page 7
WPROCorganization
Weaknesses

Fig. 1.8 page 10
PCPbusinessPhases

Fig. 1.9 page 10
PCPdesignPhases

Fig. 1.10 page 11
PCPphasesProsAndCons

Fig. 1.11 page 12
PCPcharacteristics

Fig. 1.12 page 12
FBdevelopmentModels

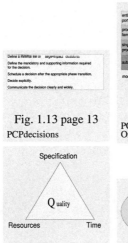

Fig. 1.13 page 13
PCPdecisions

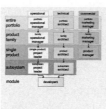

Fig. 1.14 page 14
PCPoperational
Organization

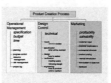

Fig. 1.15 page 15
PCPdecomposition

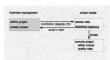

Fig. 1.16 page 17
PCPoperationalGame

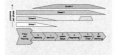

Fig. 1.17 page 17
PCPoperationalTriangle

Fig. 1.18 page 18
PCPconcentricTeams

Fig. 1.19 page 19
FBdeviationWithout
Feedback

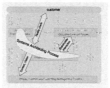

Fig. 1.20 page 20
LWAfeedback

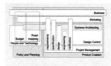

Fig. 1.21 page 21
FBschoolsOf
ArchitecturesPresence

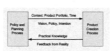

Fig. 1.22 page 21
FBphasesTheoreticalVs
Practice

Fig. 1.23 page 21
FBperPhase

Fig. 1.24 page 23
SAPprocessSimplified

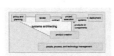

Fig. 1.25 page 23
SAPprocessMap

Fig. 1.26 page 23
SAPcouplingPPPtoPCP

Fig. 1.27 page 26
BAOPprojectVsProduct

Fig. 1.28 page 27
BAOPevolution

Fig. 1.29 page 27
PPSprojectProcess

Fig. 1.30 page 28
PPSservicesPhone
Example

Fig. 1.31 page 28
PPSservicesOperational
Model

Fig. 1.32 page 29
PPSsystemOfSystems

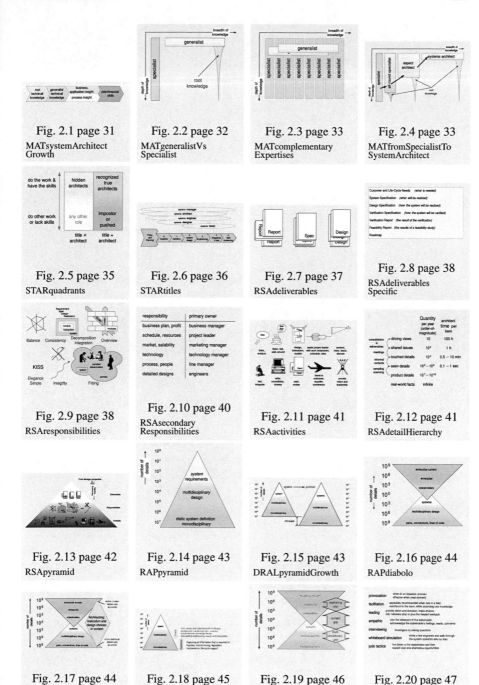

Fig. 2.1 page 31
MATsystemArchitect
Growth

Fig. 2.2 page 32
MATgeneralistVs
Specialist

Fig. 2.3 page 33
MATcomplementary
Expertises

Fig. 2.4 page 33
MATfromSpecialistTo
SystemArchitect

Fig. 2.5 page 35
STARquadrants

Fig. 2.6 page 36
STARtitles

Fig. 2.7 page 37
RSAdeliverables

Fig. 2.8 page 38
RSAdeliverables
Specific

Fig. 2.9 page 38
RSAresponsibilities

Fig. 2.10 page 40
RSAsecondary
Responsibilities

Fig. 2.11 page 41
RSAactivities

Fig. 2.12 page 41
RSAdetailHierarchy

Fig. 2.13 page 42
RSApyramid

Fig. 2.14 page 43
RAPpyramid

Fig. 2.15 page 43
DRALpyramidGrowth

Fig. 2.16 page 44
RAPdiabolo

Fig. 2.17 page 44
LAWFdiabolo

Fig. 2.18 page 45
DRALdesignEngineering

Fig. 2.19 page 46
DRALgaps

Fig. 2.20 page 47
ASstyles

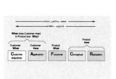

Fig. 3.1 page 52
CAFCRannotated

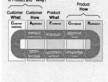

Fig. 3.2 page 52
MSintegratingCAFCR

Fig. 3.3 page 53
CAFCRrecursion

Fig. 3.4 page 53
BCAFCRwhoIsTheCustomer

Fig. 3.5 page 54
BCAFCRplusLifeCycle

Fig. 3.6 page 55
REQwhatWhatHow

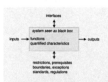

Fig. 3.7 page 56
REQsystemAsBlackBox

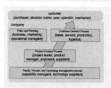

Fig. 3.8 page 57
REQstakeholders

Fig. 3.9 page 58
REQrequirementsFor
RequirementsAll

Fig. 3.10 page 59
COVmotorwayManagement
KeyDrivers

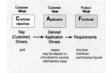

Fig. 3.11 page 60
REQfromDriverTo
Requirement

Fig. 3.12 page 61
TCAFkeyDriverSubmethod

Fig. 3.13 page 62
TCAFkeyDriver
Recommendations

Fig. 3.14 page 64
REQviewpoints

Fig. 3.15 page 65
REQselection

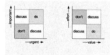

Fig. 3.16 page 65
REQqualitative
SelectionMatrix

Fig. 3.17 page 66
MPBAvalueCriteria

Fig. 4.1 page 70
TBSAtoolsMap

Fig. 4.2 page 74
KDAWStools

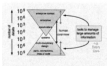

Fig. 4.3 page 75
KDAWStoolsDiabolo

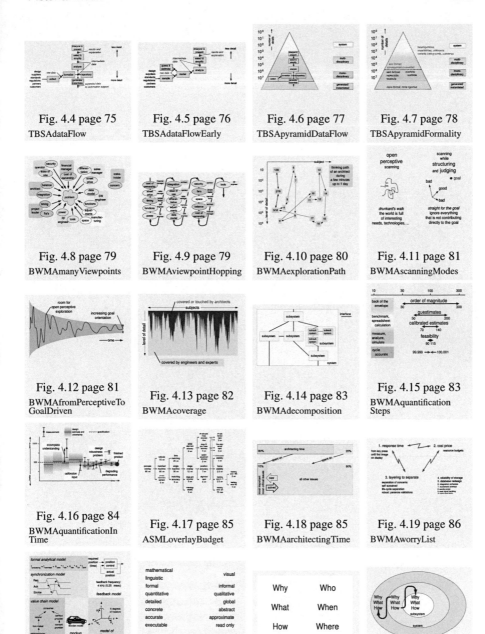

Fig. 4.4 page 75
TBSAdataFlow

Fig. 4.5 page 76
TBSAdataFlowEarly

Fig. 4.6 page 77
TBSApyramidDataFlow

Fig. 4.7 page 78
TBSApyramidFormality

Fig. 4.8 page 79
BWMAmanyViewpoints

Fig. 4.9 page 79
BWMAviewpointHopping

Fig. 4.10 page 80
BWMAexplorationPath

Fig. 4.11 page 81
BWMAscanningModes

Fig. 4.12 page 81
BWMAfromPerceptiveTo
GoalDriven

Fig. 4.13 page 82
BWMAcoverage

Fig. 4.14 page 83
BWMAdecomposition

Fig. 4.15 page 83
BWMAquantification
Steps

Fig. 4.16 page 84
BWMAquantificationIn
Time

Fig. 4.17 page 85
ASMLoverlayBudget

Fig. 4.18 page 85
BWMAarchitectingTime

Fig. 4.19 page 86
BWMAworryList

Fig. 4.20 page 87
BWMAmodelExamples

Fig. 4.21 page 87
BWMAmodelTypes

Fig. 4.22 page 88
BWMAquestions

Fig. 4.23 page 88
BWMArecursionWWH

Fig. 4.24 page 89
TORdecisionFlow

Fig. 4.25 page 90
TORmultiple
Propositions

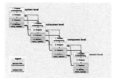

Fig. 4.26 page 91
TORrecursion

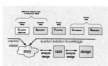

Fig. 4.27 page 92
SHTfromStoryToDesign

Fig. 4.28 page 92
SHTexampleStoryLayout

Fig. 4.29 page 94
SHTcriterionsList

Fig. 4.30 page 96
SHTexamplePortrait

Fig. 4.31 page 97
SHTexampleCriteria

Fig. 5.1 page 99
BSMMbasicConcepts

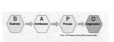

Fig. 5.2 page 100
BSMMbapoFramework

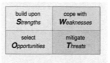

Fig. 5.3 page 101
BSMMmethodSWOT

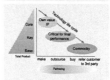

Fig. 5.4 page 101
SSScoreKeyBase

Fig. 5.5 page 102
BSMMbusinessModels

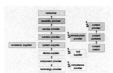

Fig. 5.6 page 103
BSMMvalueChain

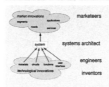

Fig. 5.7 page 103
RIIcontributors

Fig. 5.8 page 105
RSProadmapStructure

Fig. 5.9 page 105
ROADdeliverables

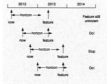

Fig. 5.10 page 106
ROADonOffManagement

Fig. 5.11 page 107
ROADanalogManagement

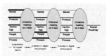

Fig. 5.12 page 108
ROADbursts

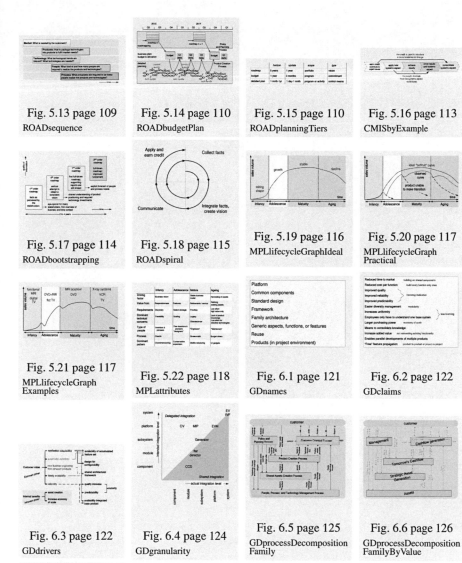

Fig. 5.13 page 109
ROADsequence

Fig. 5.14 page 110
ROADbudgetPlan

Fig. 5.15 page 110
ROADplanningTiers

Fig. 5.16 page 113
CMISbyExample

Fig. 5.17 page 114
ROADbootstrapping

Fig. 5.18 page 115
ROADspiral

Fig. 5.19 page 116
MPLlifecycleGraphIdeal

Fig. 5.20 page 117
MPLlifecycleGraph
Practical

Fig. 5.21 page 117
MPLlifecycleGraph
Examples

Fig. 5.22 page 118
MPLattributes

Fig. 6.1 page 121
GDnames

Fig. 6.2 page 122
GDclaims

Fig. 6.3 page 122
GDdrivers

Fig. 6.4 page 124
GDgranularity

Fig. 6.5 page 125
GDprocessDecomposition
Family

Fig. 6.6 page 126
GDprocessDecomposition
FamilyByValue

Fig. 6.7 page 126
GDprocessDecomposition
FamilyPlusFlow

Fig. 6.8 page 127
GDoperational
Organization

Fig. 6.9 page 129
SWRreuseModels

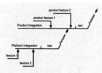

Fig. 6.10 page 130
GDpropagationDelay

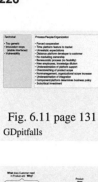

Fig. 6.11 page 131
GDpitfalls

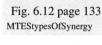

Fig. 6.12 page 133
MTEStypesOfSynergy

Fig. 6.13 page 134
EPLOapproach

Fig. 6.14 page 135
MPBAexplore

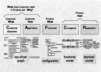

Fig. 6.15 page 135
MPBAproductMarket

Fig. 6.16 page 136
EPLOworkFlowAnalysis

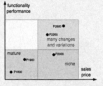

Fig. 6.17 page 137
MPBAproductMarketMap

Fig. 6.18 page 137
MPBAfeaturesTechnology

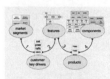

Fig. 6.19 page 138
MPBAmapping

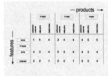

Fig. 6.20 page 139
PFproductFeatureMap
WithNumbers

Fig. 6.21 page 139
HMPAprojectsCase

Fig. 7.1 page 141
AASPonBusiness
Decomposition

Fig. 7.2 page 142
AASPflow

Fig. 7.3 page 143
DGdocumentationRoles

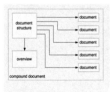

Fig. 7.4 page 146
DGcompoundDocument

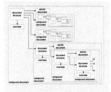

Fig. 7.5 page 147
DGdocumentRecursion

Fig. 7.6 page 148
DGpayload

Fig. 7.7 page 153
LEANoverviewA3

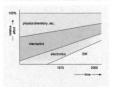

Fig. 8.1 page 156
RSWrelativeEffort

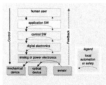

Fig. 8.2 page 156
RSWcontrolHierarchy

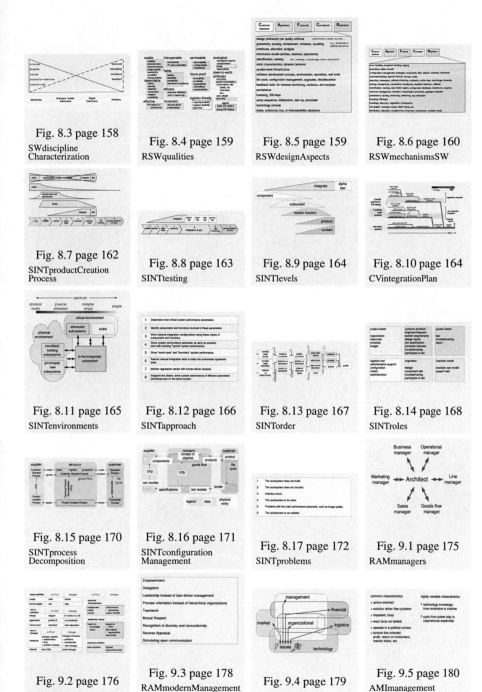

Fig. 8.3 page 158
SWdiscipline
Characterization

Fig. 8.4 page 159
RSWqualities

Fig. 8.5 page 159
RSWdesignAspects

Fig. 8.6 page 160
RSWmechanismsSW

Fig. 8.7 page 162
SINTproductCreation
Process

Fig. 8.8 page 163
SINTtesting

Fig. 8.9 page 164
SINTlevels

Fig. 8.10 page 164
CVintegrationPlan

Fig. 8.11 page 165
SINTenvironments

Fig. 8.12 page 166
SINTapproach

Fig. 8.13 page 167
SINTorder

Fig. 8.14 page 168
SINTroles

Fig. 8.15 page 170
SINTprocess
Decomposition

Fig. 8.16 page 171
SINTconfiguration
Management

Fig. 8.17 page 172
SINTproblems

Fig. 9.1 page 175
RAMmanagers

Fig. 9.2 page 176
RAMcomparison

Fig. 9.3 page 178
RAMmodernManagement
Techniques

Fig. 9.4 page 179
AMIintroduction

Fig. 9.5 page 180
AMImanagement
Characteristics

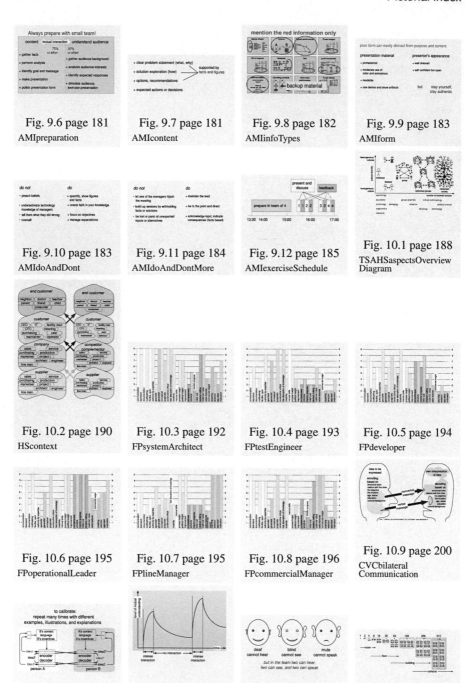

Fig. 9.6 page 181
AMIpreparation

Fig. 9.7 page 181
AMIcontent

Fig. 9.8 page 182
AMIinfoTypes

Fig. 9.9 page 183
AMIform

Fig. 9.10 page 183
AMIdoAndDont

Fig. 9.11 page 184
AMIdoAndDontMore

Fig. 9.12 page 185
AMIexerciseSchedule

Fig. 10.1 page 188
TSAHSaspectsOverview
Diagram

Fig. 10.2 page 190
HScontext

Fig. 10.3 page 192
FPsystemArchitect

Fig. 10.4 page 193
FPtestEngineer

Fig. 10.5 page 194
FPdeveloper

Fig. 10.6 page 195
FPoperationalLeader

Fig. 10.7 page 195
FPlineManager

Fig. 10.8 page 196
FPcommercialManager

Fig. 10.9 page 200
CVCbilateral
Communication

Fig. 10.10 page 200
CVCcodingCalibration

Fig. 10.11 page 201
CVCunderstandingInTime

Fig. 10.12 page 202
HSTWthreeApes

Fig. 10.13 page 202
BLOATorganization

Fig. 10.14 page 203
HSTWsimpleTeamModel

Fig. 10.15 page 203
HSTWhierarchicalModel

Fig. 10.16 page 204
HSTWbelbinRoles

Fig. 10.17 page 204
HSTWSixThinkingHats

Fig. 10.18 page 205
HSTWmyersBriggs

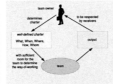

Fig. 10.19 page 206
HSTWcharter

Fig. 10.20 page 207
HSTWwarRoom

Fig. 10.21 page 207
HSTWfragmentation

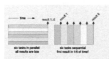

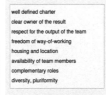

Fig. 10.22 page 208
HSTWconcurrency

Fig. 10.23 page 208
HSTWmultipleTeams

Fig. 10.24 page 209
HSTWcriticalSuccess
Factors

Fig. 11.1 page 211
RASAscope

Fig. 11.2 page 212
RASAcycle

Index

T - #0121 - 101024 - C0 - 234/156/15 [17] - CB - 9781439847626 - Gloss Lamination